UNRAVELLING THE MIND OF GOD

Unravelling the Mind of God

Mysteries at the frontier of science

Robert Matthews

Quality Paperbacks Direct
London

This edition published 1992
by QPD by arrangement with Virgin Publishing Ltd
338 Ladbroke Grove
London W10 5AH

A catalogue record for this title is available from the British Library
CN 3698

Printed and bound in Great Britain by
Mackays of Chatham PLC,
Chatham, Kent

Contents

Preface

THERE ARE A GREAT MANY deep mysteries in science. This may come as a surprise to those who have read some other popular accounts of science, which give the impression that Einstein and a few others pretty well tied everything up years ago.

This book is an invidious selection of some of these outstanding mysteries, the efforts that have been made to solve them and the questions that remain. Some of the questions are literally cosmic in magnitude: how did the universe begin? How will it end? Others are surprisingly down to earth: why do we age?

My guiding principle in making the selection has been the Gossip Test, whose origin can be traced back to Francis Crick, co-discoverer of the structure of the 'life molecule' DNA. At the heart of the test is a question: can you imagine two friends talking about a particular Outstanding Mystery after work?

Clearly, the test must be applied with care. Professional scientists who have burrowed deep into their subject find esoteric discoveries incredibly exciting and worth bending the ear of all who will listen. And indeed they often are – once one knows how the discovery fits into the grand scheme of things.

And this has been my aim with this book: to provide a mini grand tour of some of the most exciting areas of science, so that the reader with no scientific background at all can see what all the fuss is about.

To do this, I have shamelessly adopted the Bluffers' Guide approach to each broad area covered by this book. My aim has been to give readers what they need to know to get a feel for how the particular field got to where it is now, and for the personalities who put it there. Each chapter then moves on to look at the Outstanding Mysteries that are the focus of research now.

At all times I have tried to avoid the science textbook approach.

Textbooks have, I believe, played a key role in one of the greatest intellectual tragedies of our time: the Dulling of Science. It is a deep irony that at a time when the impact of scientific advance on our lives has never been greater, general knowledge of science is so dismal. Somehow, a subject that every young child finds fascinating has been turned into a dreary subject which is widely loathed – and feared.

Professional-level science is undoubtedly difficult. If you want to really understand what happened in the very earliest moments of the universe, you do have to be pretty good at mathematics – perhaps the only subject most people loathe and fear even more than science. But few people ever want to acquire that level of knowledge.

This does not seem to have dawned on the architects of science curricula in schools and universities. In Britain, things seem to be getting better, but for years science education appeared to be in the hands of academics who had forgotten (if they ever knew) the real excitement of science. Latter-day Gradgrinds, they turned science into a minority interest for fact fetishists.

Professional scientists themselves must also carry some of the responsibility for the parlous state of scientific literacy in Britain and America. With some laudable exceptions, they have been less than forthcoming about the excitement that can be had unravelling the mysteries of Nature.

There remains, however, an ineluctable problem facing anyone attempting to popularise science: just how much detail *does* one leave in? Scientists have of course discovered an awful lot about many things. Most of this is fine detail – often very fine, very complex and very dull detail. Many scientists are fascinated by fine detail. Some of them even put lots of it into their 'popular' accounts. All of it is studiously ignored in this book.

Many scientists will think I have drawn the line at too simple or simplistic a level. Too bad – I'm not writing for the already expert. Others might find the going rather confusing or hard in parts. It may then be comforting to know that even Einstein would not accept some of the ideas in this book. My advice to anyone getting bogged down is – skip. Move on to another chapter, even. The chapters are reasonably self-contained, and the index should help fill in any details you miss in the jump.

Ultimately, why should anyone know anything about science? Certainly a knowledge of theories about the origin of life isn't likely to win you much cachet at a dinner party. Utilitarian arguments can be used to justify some scientific knowledge. Chemistry has led to the development of a host of compounds, from anticancer drugs to biodegradable plastics. Physics has shaped modern society through such fields as electromagnetism and aerodynamics. Even particle physics, which some scientists themselves regard as an expensive indulgence, has produced spin-offs like positron emission tomography scanners in hospitals and neutron beam analysis of materials.

But such arguments are boring and overly defensive. The reason I believe everyone should take an interest in science is the reason most scientists get involved in it: science, at its most basic level, is about trying to work out why the universe is as it is.

Writing in *A Brief History of Time*, perhaps the most famous popularisation of science ever written, Professor Stephen Hawking puts the aim of science in grander terms: he calls it an attempt to know the mind of God. It's time everyone had access to what we have learnt so far.

Acknowledgements

ONE OF THE PRIVILEGES of being a science correspondent is the access it brings to those working at the very forefront of research. It would have been impossible to write this book without being able to draw on their knowledge and insights, both directly and through their own writings. I have benefited from the advice of the many scientists I have talked to as a result of covering science issues in their areas. Space does not permit a full listing; however, I should particularly like to thank Professor John Barrow, Professor Geoff Brown, Dr Graham Currie, Professor Angus Dalgleish, Professor Paul Davies, Professor George Kalmus and Dr Jonathan Powers for generously agreeing to read through early drafts of various chapters and giving me much-needed guidance and advice.

On specific matters, I have greatly benefited from conversations with Professor David Bohm, Dr Terry Clark, Dr Victor Clube, Dr John Ellis, Professor George Efstathiou, Dr Joe Farman, Professor Michael Green, Professor Barry Hall, Dr Charles Hardingham, Professor Chris Isham, Dr Ian Jackson, Dr John Jackson, Dr Ian Moss, Dr Peter O'Hare, Dr Arno Penzias, Professor Martin Rees, Professor John Schwarz, Professor Ian Stewart, Dr John Sulston, and Professor Edward Witten. Of course, any errors that remain are entirely the result of my own misunderstandings and inadequacies.

My final debt of gratitude is to Fiona Bacon, whose tolerance runs far beyond the call of marriage.

1 A Tour of the Creation

I N THIS BOOK, we shall be exploring the frontiers of what for many readers will be unfamiliar territory. We shall venture deep into the cosmos to seek answers to questions about the origin and fate of the universe. We shall be probing to the very heart of matter, down to the strange realm of the quantum where subatomic particles turn into ghosts. On the way we shall encounter living cells and the viruses that prey on them, dinosaur-killing comets and exploding stars.

When travelling through unfamiliar territory, a map boosts one's confidence of reaching the final destination. This brief opening chapter will, I hope, serve as a map of what is to come.

A powerful simplification

The sheer scope of modern science beggars the imagination. Scientists routinely study galaxies billions of light years distant, and particles a million-billionth of a metre across; they talk of life on Earth hundreds of millions of years ago and of events in the first millionth of a second after the Big Bang that started everything off.

So vast a range of phenomena demands a simple way of describing their essential properties, lest our descriptions become swamped by long strings of noughts. For example, written out in full, the age of the universe is thought to be about 15,000,000,000 years. A hydrogen atom, the most common in the universe, weighs about 0.0000000000000000000000002 kilograms. Fortunately, a more convenient way of expressing such things does exist. It is known as *exponential notation*, and is used by scientists specifically to get around the need to write down all these noughts. The noughts are accounted for by using 'powers of ten', where the 'power' is equal to the number of noughts involved. Thus one million, written in full as 1,000,000, becomes 'ten to the power six', written as 10^6. A billion (and in this book we use the widely accepted American version of a billion as being a thousand millions) written out in full is 1,000,000,000, but in the exponential notation is just 10^9. Thus the age of the universe can be written as 15×10^9 years.

For very small numbers the same trick is used, but this time with a minus sign in front of the power. Thus one millionth, which in full would be 0.000001, is instead written as 10^{-6}. One billionth, written 0.000000001, becomes 10^{-9}. For example, the mass of the hydrogen atom is 2×10^{-27} kilograms. To complete the picture, 1 is simply 10^0 (there are no noughts involved), and 10 is 10^1.

A ride on Briggs's Elevator

As well as giving us a compact way of writing down both immensely large and extremely small numbers, the concept of powers of ten provides us with a convenient foundation for a tour of the Creation. The American chemist Robert Shapiro was the first, to my knowledge, to come up with the idea of a magical elevator which can take us not merely to different levels, but to different *scales* of phenomena. Each 'floor' on the control panel of this elevator thus represents, not a different height above or below street level, but a different scale of phenomena – greater or smaller than those of ordinary life. Thus the numbers on control panel buttons represent new powers of ten.

I propose to call this magical device Briggs's Elevator, as a tribute to the seventeenth-century English mathematician Henry Briggs who popularised the notion that any number can be expressed in terms of powers of ten.

The ground floor, Level 0, corresponds to the scale 10^0, i.e. 1. It is full of everyday phenomena: people, cars, cats and dogs. Pressing the button marked Level 1 takes us up to Level 10^1 – a scale ten times larger. Looking out from the elevator, though, we notice few differences – except that we can see rather further than before. Let us therefore press for Level 4. Now we find that the scale has been changed by a factor of 10,000 (10^4). For every metre we could see before, we can now see ten kilometres (10^4 metres). We are able to look down on our home town and just about make out where we live in it.

Briggs's Elevator takes us on a tour of the Creation

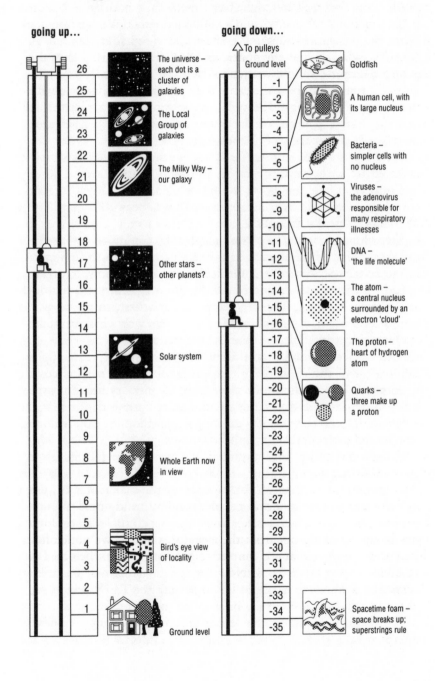

going up...

Floor		
26		The universe – each dot is a cluster of galaxies
25		
24		The Local Group of galaxies
23		
22		The Milky Way – our galaxy
21		
20		
19		
18		
17		Other stars – other planets?
16		
15		
14		
13		Solar system
12		
11		
10		
9		
8		Whole Earth now in view
7		
6		
5		
4		Bird's eye view of locality
3		
2		
1		
Ground level		

going down...

To pulleys
Ground level

Floor		
-1		Goldfish
-2		A human cell, with its large nucleus
-3		
-4		
-5		Bacteria – simpler cells with no nucleus
-6		
-7		
-8		Viruses – the adenovirus responsible for many respiratory illnesses
-9		
-10		
-11		DNA – 'the life molecule'
-12		
-13		
-14		The atom – a central nucleus surrounded by an electron 'cloud'
-15		
-16		
-17		The proton – heart of hydrogen atom
-18		
-19		
-20		Quarks – three make up a proton
-21		
-22		
-23		
-24		
-25		
-26		
-27		
-28		
-29		
-30		
-31		
-32		
-33		
-34		Spacetime foam – space breaks up; superstrings rule
-35		

Pushing the button marked '8', Briggs's Elevator takes us up to a much more impressive sight. Every metre has become a hundred million metres – 10^8 metres or 100,000 kilometres. We can now look down on the entire planet. We share the views first seen by Yuri Gagarin, who orbited the Earth in 1961. For countless centuries, people thought – sensibly enough – that the colourful ball of multi-hued continents, blue seas and white clouds now in view must be the centre of all things. The sun's steady movement across the sky, and the stately passage of the planets through the constellations, served only to confirm this comforting picture.

But when we move up to Level 13, we see the dangers of being deceived by such obvious facts as the sun's motion across the sky 'proving' that the sun goes around the Earth. In 1990, the first complete photograph taken from Level 13 was beamed back to Earth by the space probe Voyager 2, after a thirteen-year journey past the furthest-flung planets. The picture graphically confirmed what radical thinkers, from Aristarchus of Samos 2,300 years ago to Copernicus in the sixteenth century and Galileo in the seventeenth, had maintained: the Earth is but one of a collection of planets, all of which are orbiting the sun. The sun's apparent motion across the sky is simply an illusion, the result of the Earth's daily rotation on its axis.

On the Voyager picture, the Earth appears as a tiny blue dot, delicate and vulnerable in the abyss of space. Indeed, it is only the thin veneer of the atmosphere that protects us from myriad cosmic threats that constantly assault the Earth. Its mere presence provides a shield against the thousands of tonnes of cosmic debris that the Earth ploughs through on its year-long trip around the sun, remnants of wrecked asteroids and long-dead comets.

But as we shall learn in Chapter 3, the atmosphere has not always succeeded in preventing catastrophe. In 1980, American scientists announced that they had found a layer of peculiar material buried in the Earth's crust at a depth corresponding to a period about 65 million years ago. That epoch was already well known to geologists for another reason – it marks the mysterious end of the 130-million-year reign of the seemingly invincible dinosaurs. The chemical composition of that layer of material strongly suggests that 65 million years ago a huge meteorite or comet struck the Earth, triggering a global conflagration that may have killed the dinosaurs.

The disturbing fact is that there are still very large chunks of interplanetary rock orbiting the sun, uncomfortably close to the

Earth. As recently as 1908, a huge area of Siberia was devastated by what appears to have been part of a comet that managed to get through most of the atmosphere before disintegrating. Whether we may share the same Damoclean fate as the dinosaurs – and, we now know, countless other species over the Earth's history – is an Outstanding Mystery explored in this book.

Environmental assaults

Scientists are deeply concerned about the ability of the atmosphere to protect us from two other threats to life: dangerous radiation from the sun and the bitter cold of space. About 15 to 30 km above the Earth lies the so-called ozone layer. Ozone, a noxious gas closely related to oxygen, absorbs much of the ultraviolet radiation that comes from the sun. Ultraviolet rays are responsible for giving us a tan on the beach. But in excessive quantities they cause sunburn, and even melanomas (skin cancers). Man-made compounds known as chlorofluorocarbons (CFCs), once used widely as aerosol propellants and still used by some industries, destroy ozone. After decades of use they have seeped into the atmosphere, and are now attacking the ozone layer above our heads. Holes have started to form in this layer, allowing more UV radiation to pour through. Urgent action is now being taken to stop an explosion of skin-cancer cases and a myriad other problems.

Yet the problems caused by a punctured ozone layer pale in comparison to the threat of global warming.

Surprisingly enough, were it not for traces of certain vapours and gases in the atmosphere (primarily water vapour and carbon dioxide) the Earth would be plunged to a temperature of about $-18°C$: the sun's heat is just too feeble by itself to sustain a higher temperature. But these atmospheric gases – known as greenhouse gases – have the ability to trap the sun's warmth, and boost the final temperature by more than 33 degrees to a much more acceptable $15°C$ average.

However, since the Industrial Revolution, huge increases in the burning of fossil fuels like coal have dramatically increased the level of carbon dioxide in the atmosphere. The levels of other greenhouse gases like methane, produced by rotting organic material, have also grown. The result is that the atmosphere may now be doing too good a job of trapping the sun's heat. There is evidence that the temperature of the Earth is slowly rising. Over the next fifty years, unless we take avoiding action, scientists think the Earth will warm by up to 3°C. This may not sound very dramatic, but such is the delicate balance of our weather systems that it is enough to raise the spectre of dramatic changes in agriculture and land use which could have catastrophic results. Wars over scarce water reserves is just one of the threats now being talked of. Many scientists think that the time has come to take action on atmospheric pollution to prevent the balance being disturbed further.

Our ability to predict the nature of these changes is of vital import-ance if we are to avoid economic disaster from unforeseen effects of the warming. The prediction of natural events is now the subject of intense research, and involves a relatively new and intriguing idea: chaos theory. Some of the implications of this are explored in Chapter 3.

Let us now leave the Earth behind, and take Briggs's Elevator to Level 17. Up here, we discover that the sun may not be alone in having a family of planets. Rings of dust around other stars on our cosmic doorstep suggest that planet formation may not be all that rare. But whether life can also exist elsewhere in our galaxy is an Outstanding Mystery also examined in Chapter 3.

Moving up to Level 22, we find that the sun, despite its vast importance to us, has become just an insignificant star among many billions making up a huge spiral galaxy. The Milky Way, the faint streak of phosphorescence that cuts across the night sky, is part of one of the spiral arms of that galaxy. Our solar system lies about two thirds of the way out from the centre, and takes about 200 million years to complete a single orbit. The Earth has thus made fewer than 25 trips around the galaxy since its creation.

Looking from our elevator at Level 24, we see that our galaxy is a member of a small cluster of others, a few also spirals, but mainly curiously compact, elliptical-shaped conglomerations of stars. We are looking at the so-called Local Group of galaxies. By Level 25 we see that it too is one of untold millions of clusters spread through the vastness of space.

The final button on the control panel of Briggs's Elevator is 26. When we press it, we can finally view the whole of the observable universe in one huge panorama. Individual galaxies can no longer be discerned. Our solar system is utterly lost to view. All we can see are smudges of light representing the very largest clusters of galaxies. All are racing away from one another, propelled by the impetus of the

Big Bang explosion that gave birth to the universe about fifteen billion years ago. Whether those clusters will continue their headlong race from one another depends on something we cannot see as yet detect: mysterious dark matter lurking in and between the galaxies, subtly exerting its influence on the whole cosmos. If enough dark matter exists, the expansion of the universe will eventually come to a halt and then reverse. The clusters of galaxies we now see will start to come together again, merging and finally colliding, heading for a cataclysmic Big Crunch billions of years from now. Will such a set of events come to pass? How did the universe start in the first place? Tentative answers to these Outstanding Mysteries will come in Chapter 6.

With the whole of the observable universe now filling the viewing window of the Elevator, we can go no further. But our tour of the Creation is far from finished. We can continue our journey by going down below 'ground level' to scales far beneath those of everyday experience. What we will meet there will seem much stranger still.

Life and death under the microscope

At Level -1, where the natural scale of things is not the metre but 0.1 of a metre, we are still in familiar territory. Let us therefore drop straight down to Level -5, where objects 0.00001 (10^{-5}) of a metre across come clearly into view. At this level we can see individual human cells – 10,000 of which would barely cover the head of a pin. From the Elevator's window, the intricate workings of these 'life factories' can be made out. A double-walled outer membrane marks the boundary of the 'factory', letting in nutrients extracted from food, and keeping out invaders such as bacteria. Oval-shaped objects called mitochondria generate power from the nutrients and keep the chemical processes essential to life chugging away.

Go down another level, to -6, and the part of the cell from which the whole operation is controlled – the cell nucleus – comes into view. This contains the chemical blueprint for everything the cell does. The blueprint is in the form of strands of a complex, helix-shaped chemical molecule known as deoxyribose nucleic acid – DNA. In humans, the complete DNA blueprint is divided among 23 pairs of 'files' known as chromosomes: curious, mainly x-shaped objects, which, when stained, reveal a series of bands across them. Incredibly, each of these minuscule chromosomes contains about four centimetres or so of DNA; thus, if the total chemical blueprint packed into each tiny human cell were teased out of the nucleus and unravelled, it would stretch for almost two metres.

Virtually every cell in the body contains these chromosomes. They all have the same instructions, because they are all descendants from one single cell: the fertilised egg. A sperm carrying 23 chromosomes combines at conception with an egg carrying the other 23. The fertilised egg then undergoes a complex series of divisions into two cells, then four, then eight and so on, becoming more and more complex. As development continues, the cells start to take on their own individual roles. They do this by chemically 'reading' only certain parts of the total DNA blueprint, known as genes. Thus some cells read only those genes in the blueprint that will enable them to become, say, kidney cells. Others become skin cells. The whole process normally takes place in perfect harmony, producing a foetus, then a living baby, a child and finally an adult. Like so much in biology, the whole process is seemingly miraculous. How can such stunning complexity be produced from such simple beginnings? How, in short, does a single cell far smaller than the dot of this letter 'i' produce a sentient being? Many of the processes involved remain mysterious, and constitute one of the most fascinating of all the Outstanding Mysteries. We shall look at it in Chapter 2.

Down at Level −6, we see that human cells are far from being the only living things around at this scale. We see bacteria − organisms about the same size as the nucleus of one of our cells, but quite capable of making us extremely ill, or even killing us. They come in a variety of shapes, their names associated with some of the worst scourges of man: the spiral-shaped spirochaetes, responsible for syphilis; the rod-shaped bacilli that wiped out much of the population of mediaeval Europe in the Black Death; and the ball-shaped streptococci, killer of so many children by scarlet fever in the last century.

Fortunately, bacteria are no longer the threat they were. Antibiotics like penicillin are one of the great successes of modern medicine. But scientists have had far less success with the deadly entities we only glimpse down here at Level −6: viruses. If we travel down to Level −8, these bizarre objects now loom into view. The largest are the so-called pox viruses. Finding a way into a healthy cell, they hijack the cell's chemical-making system. The cell is forced into making compounds that suit the virus, not itself.

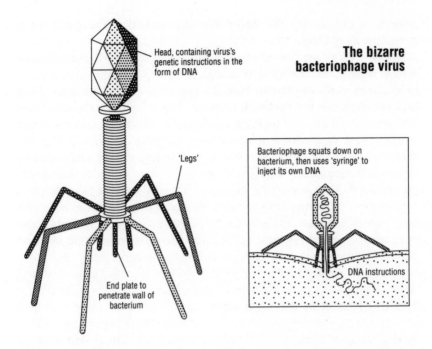

Head, containing virus's genetic instructions in the form of DNA

The bizarre bacteriophage virus

'Legs'

Bacteriophage squats down on bacterium, then uses 'syringe' to inject its own DNA

End plate to penetrate wall of bacterium

DNA instructions

Infection with the pox viruses results in diseases like smallpox. Fortunately, a huge programme of vaccination by the World Health Organisation during the 1970s and 1980s has more or less eradicated smallpox from our planet. But that has been one of the few successes against these molecular pests. There are a myriad other types of virus that threaten us, many of which, with their intricate, multifaceted bodies, appear deceptively attractive from our viewing window. The adenovirus, responsible for many respiratory ailments, looks like some minuscule space probe, with long aerial-like proteins sticking out from a twenty-sided central mass. The bacteriophage virus, so called because it preys on bacteria, looks as though it has been designed by a race of aliens (see above). It has a multifaceted head connected by a long neck to spindly legs that enable it to clamp on to the side of a bacterium and inject its chemical demands directly into the heart of a living cell.

And now we meet the most notorious virus of them all – the Human Immunodeficiency Virus, HIV, otherwise known as the AIDS virus. Its bland, roughly spherical shape belies the appalling havoc it is now wreaking in the world. Will it ever be defeated? In Chapter

2 we look at some of the many strategies scientists are drawing up in their war against HIV.

Smallest of all are the picornaviruses – just 2×10^{-8} metres or so across. It seems incredible that we cannot prevent so tiny an organism from causing misery, but such is the case: one type of picornaviruse is responsible for the common cold.

Our relative lack of success in fighting viral infection becomes clearer when we go down another level, to Level -9, and peer inside at the heart of the virus. We find that they contain the same sort of chemical blueprint we find inside our own cells. We can discern the chemical DNA in some, and in others its cousin molecule, RNA.

Scientists think that the reason viruses have similarities to the workings of animal cells is because the viruses are, in fact, 'on the run'. And therein lies one of the key problems in tackling viral diseases: killing the virus without harming the cells.

Into the Realm of the Quantum

Taking Briggs's Elevator to Level -10, we notice things are starting to become a little strange. It is difficult to say what exactly is happening, but somehow, down here at the level of atoms and molecules, things are becoming a little, well, hazy. This is the first warning we have got that we are entering the Realm of the Quantum.

In the late nineteenth and early twentieth century, scientists began to discover that at the level of atoms commonsense ideas about the nature of matter start to break down. As children we are often taught to think of atoms as miniature solar systems, with electrons whizzing like planets around a central 'sun' – the atomic nucleus. From our Elevator, we discover that such a picture is *very* wide of the mark. For a start, we can see neither electrons nor the nucleus. In the Realm of the Quantum, as we shall learn in Chapter 4, one must get used to disbelieving one's own senses, one's own intuition. Particles are not like the billiard balls. They have become ghostlike mirages, bound to one another in ways that defy description.

Scientists have used huge particle accelerators to probe down deep to where we are now in the Elevator, in a search for the ultimate constituents of matter. And they have a warning for us: to peer out of the elevator at such depths is to risk one's sanity. Let us heed their warning, but press on downwards nonetheless, closing the shutters of the Elevator's window and relying on the Elevator's on-board computer to tell us what is happening outside.

As we descend through Levels -11, -12, -13 and -14, we encounter nothing new: the atoms that make up all matter are, curiously enough, almost entirely empty space. At Level -15 we finally encounter the proton, the positively charged particle which binds the electrons in their orbits around the atom. Shortly afterwards the Elevator's computer tells us that we have reached the level of the quarks – which reside inside the proton and similar particles. Next we reach the scale of the tiny electron, lightest of the particles that make up normal matter.

But by about Level -18, this flurry of encounters is over. We seem to have come to the end of our journey. So where does our Elevator's lift shaft end? We continue our descent through the -20s and into the -30s, our computer remaining silent about conditions outside. By now we are far, far from the world we know, and deep in the Realm of the Quantum.

Now we are at Level -34. And suddenly whatever horrors there are outside the Elevator are beginning to make themselves felt inside. The sides of the Elevator are starting to buckle. The computer seems to have broken down – its screen is covered with gibberish. In a brief moment of lucidity it flashes up one last message: 'WARNING: Spacetime Foam' and goes dead. We have, at last, reached our journey's end. At Level -35, we have reached a scale so small that even tiny electrons appear to us to be the size of entire galaxies. Space and time, which at our human scale of things appear continuous, have down here dissolved into 'spacetime foam'. Thrashing around in the chaos outside the Elevator may be superstrings, multidimensional inhabitants of this wild province whose writhings could perhaps be the source of all the particles in the universe, as we shall learn in Chapter 5.

If we stay down here an instant longer, we shall be engulfed by a tsunami of spacetime, and drowned in a quantum sea. Pressing the Ascend button, we hurtle back towards the world we know, up past the electrons and protons, bursting through the portals of the Realm of the Quantum into the world of viruses and bacteria, then up past living cells, tiny mites and dust. Finally we are back, safe and comfortable in our room.

Yet above and beneath us – right now – all the things we have seen from the Elevator are still out there . . .

Our brief tour of the Creation is complete. It is time to take a closer look.

2 Questions of Life and Death

THE BRUSQUE BUT BRILLIANT PHYSICIST Lord Rutherford once pronounced that there are two types of science – physics and stamp collecting. Certainly, physicists were making fundamental breakthroughs when many other sciences either did not exist or were little more than exercises in cataloguing.

But in recent years many of the best minds in physics have soared off into intellectual realms far removed from everyday experience: black holes and subatomic particles are their stock in trade. More or less contemporaneously, however, biologists have emerged from long years of cataloguing to attack questions of far greater relevance to everyday existence. How, for instance, does a single fertilised egg cell turn into a human being – a collection of billions of cells of more than a hundred different types, all performing specific tasks? As we shall see, more or less every cell has the potential to develop into any other cell in the body, so why don't they exercise that ability? Why, in short, is a liver cell not a lung cell? Come to that, how did life itself originate? The answers to any of these questions are the stuff of Nobel prizes.

The questions now facing biology (I use the term in its widest sense) may still be very basic; that, for me at least, is its great attraction. But the progress being made in answering them is being put to use in some crucial fields: the development of new drugs, the detection and prevention of genetic diseases, the search for vaccines against AIDS.

Those scientists and researchers working in biology now are living in a golden age of discovery.

How to make a living creature

Virtually all we need to know to get to grips with the deepest mysteries of the life sciences is contained in the workings of living cells – little bags of fluid about 0.01 of a millimetre across. Our bodies contain about 10^{13} cells. If they perform properly, we thrive. We can fight infection, repair wounds, grow stronger. If they misbehave, we may find ourselves in serious trouble. Cancer, AIDS and senile dementia are all the result of cells failing to do their job correctly.

The prime job of a cell, however, sounds rather mundane: it is the production of proteins. Indeed, the best way to think of a living cell is as a chemical factory, taking in raw ingredients obtained from food at one end, and churning out tailor-made proteins at the other.

Mention protein to most people, and they think of steak dinners. It is true that the major structural component of bones, tendons, ligaments and skin is a protein known as collagen. However, proteins in general can perform rather more impressive feats than enabling us to grow fat. The so-called antibodies that fight infection in our bodies are proteins. Many of the hormones which regulate aspects of our metabolism are proteins – insulin, which controls the way we use sugar as an energy source, is an example. Enzymes – compounds which promote chemical reactions essential to life – are also proteins. (Incidentally, proteins acting as enzymes can usually be recognised as such because they have names ending in -ase. Thus the enzyme in saliva which converts starch into sugars – making bread taste sweet when eaten – is called salivary amylase).

To understand life, we must understand how living cells convert the compounds they take in from food into proteins.

Anatomy of a living cell

A living cell is an astonishingly intricate and ingenious little object, but basic elements are easily summed up – see the figure overleaf for the essential layout.

Like a factory, the cell has a boundary wall which allows nutrients and certain other proteins to pass in, and newly made proteins to pass out.

At the centre of the cell we find its most important and intriguing part: a roughly circular dark patch known as the nucleus. This is the

A simplified picture of a typical human ('eukaryotic') cell

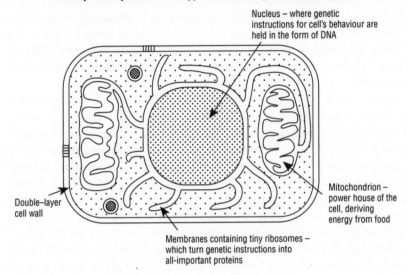

Nucleus – where genetic instructions for cell's behaviour are held in the form of DNA

Mitochondrion – power house of the cell, deriving energy from food

Double–layer cell wall

Membranes containing tiny ribosomes – which turn genetic instructions into all-important proteins

'head office' of the protein factory, from which instructions for protein manufacture ultimately issue.

The actual production line for protein manufacture lies inside the cell walls on a series of tubelike arrangements that spread out from the central region. Set along these tubes are the protein production units themselves: the ribosomes. These read the chemical instructions sent from the nucleus on how to make a protein, and turn out proteins to order.

Clearly, there must be some power source for all this activity. The 'fuel' burnt by the protein factory is the organic material contained in food. Inside the cell, objects called mitochondria take this organic material and, using enzymatic proteins, break it down into its constituents. In the process, the chemical energy stored in the material is released into the cell. But if it is not needed immediately, rather than just throw this energy away, the cell can put it in a store.

Inside the mitochondria, the food energy is trapped and stored in the form of a chemical known as ATP. In this form, the cell can call up energy if and when it needs it.

So now we have our protein factory: the cell membrane letting nutrient in, the nucleus issuing instructions on what proteins are to be made, and the ribosomes faithfully churning out the required proteins.

Protein manufacture is occurring all the time in our bodies, and in those of all other complex living creatures. If it weren't, we would drop dead.

Not all living things have such complex cells, however. Indeed, living creatures can be classed according to how sophisticated their cells are.

For example, a bacterium is made up of a single cell less than one tenth the size of an animal cell. Under the microscope, it appears to be a lot simpler in construction. There is no nucleus and no mitochondria. But anyone who has experienced a bout of food poisoning from, say, the *Salmonella enteriditis* bacterium will know that it lacks for nothing in potency. It turns out that bacteria do, in fact, have components that carry out the key functions of the nucleus and mitochondria. They also contain ribosomes, and these churn out proteins by following chemical instructions.

But despite these similarities, there is a sense in which bacteria really are simpler than animal cells, and biologists distinguish between the two cell types. Those cells – principally bacteria – which do not have a nucleus and complex cellular infrastructure are called prokaryotes, and those which do – such as human cells – are known as eukaryotes.

With that brief cell anatomy lesson out of the way, let us immediately apply it to one of the greatest mysteries modern biology has attacked to date: how does a living cell know what to do? Understand its instructions, and you have the secret to life itself.

The discovery of that secret is one of the most important and fascinating stories in the whole of science.

The search for the key to life

The first clues to what and where the instructions for life may be emerged around the 1850s. Working at the University of Berlin, a nerve-tissue expert called Robert Remak pointed out that when cells multiply to form new tissue, these cells always appear to divide themselves in two. For years, no one took much notice of this discovery. But then the influential German pathologist Rudolph Virchow realised that the idea that every cell is born from a pre-existing one could give profound new insights into the nature of disease. It became clear that it was not organs, but individual cells, which are actually affected in disease.

Soon other researchers started to look at this curious process of cell division. They noticed that it was accompanied by some odd goings-on inside the nucleus of the cell. Inside the nucleus thread-like objects – dubbed chromosomes because of their ability to take up coloured dye – were seen to come into view as the cell prepared to divide in two. What were these chromosomes doing?

By the end of the nineteenth century, chromosomes had been shown to play a key role in reproduction. It was found that the human cells specifically designed to take part in reproduction – egg cells and sperm cells – each contained 23 of these chromosomes, only half the number found in most cells in the body. It was also found that when a human egg and sperm meet and fuse at conception, the result is a cell with the usual 46 chromosomes inside its nucleus.

So what are chromosomes made of that makes them so important? By the 1930s, the answer – or, rather, two possible answers – had emerged. Chromosomes were found to contain proteins, together with a compound called deoxyribose nucleic acid, or DNA – the name of the compound reflecting its nature (acidic) and location (in the nucleus) within the cell.

At the time it seemed obvious that the proteins, rather than DNA, held the key to the secret of life. The argument was simple and elegant. Cells produce proteins, and one obvious way in which chromosomes could direct the all-important protein production of cells would be by carrying master copies – 'blueprints' – of every protein the cell ever produced.

This compelling – and completely erroneous – idea received equally compelling and erroneous support from the pronouncements of the scientist regarded in the 1930s as the greatest living expert on DNA, the other candidate for carrying the protein-making instructions. Basing his conclusions on a lifetime's research into nucleic acids, Professor Phoebus A. Levene of the Rockefeller Institute in New York declared that DNA was a small, simple and rather boring chemical. As such, it was obviously incapable of controlling anything as complex as the workings of a living cell.

When an acknowledged expert such as Levene makes a pronounce-ment the scientific community tends to listen. If it likes what it hears – for example, if what is being said backs up some other prevailing idea, and doesn't conflict with what is already known – the scientific community takes the statement on board. If, in addition, the pronouncement escapes criticism or contradiction for long enough, it can turn into an 'obvious truth'.

And so it proved with the concept of proteins as the key to life. With the world authority on the alternative – DNA – giving his own speciality the thumbs down, scientists became convinced that proteins 'must' direct the functions of the cell. If DNA did anything at all, the argument ran, it was probably something boring like providing the more complex proteins with a form of chemical scaffolding.

Ironically, even as Levene was airing his views, the experiments which were to prove him wrong were being carried out at his very own establishment.

Working with different types of the bacterium responsible for pneumonia, Oswald Avery and his co-workers at the Institute discovered that it was possible to extract from one type of the bacterium a 'transforming factor' that had the power to force another type of bacterium to take on some of its characteristics. But they also found that all the offspring of these altered bacteria had also acquired the new characteristics – they had inherited the trait from their parents.

It took Avery and his team years to establish to their own satisfaction what this 'transforming factor' was. But in 1944 they were able to announce that it was made of pure, unadulterated DNA. So much for the idea that DNA was just the scaffolding for proteins.

But the conclusion that DNA, rather than protein, is the chemical that controls cellular behaviour did not win immediate acceptance from the scientific community. There were claims that impurities in the DNA were responsible for the result. Some argued that the result was of little significance, because it applied to only one trait of only one type of living creature – and a pretty unimpressive one at that.

However, Avery's results at least got a hearing, and attracted interest – if not acquiescence – from the scientific community. The trouble with the idea that DNA held the key to life was that DNA still seemed to be such a boring compound. Analysis had shown that it was made up of just four basic compounds or 'bases', called adenine, cytosine, thymine and guanine, or A, C, T and G for short. Grand Old Man Levene had claimed that these four bases were arranged in DNA in a regular repeating pattern. Compounds stuck to the bases enable them to be joined to a chemical 'backbone' running through the DNA molecule.

Compared with the complexity of proteins, this supposedly repetitive chain of just four bases seemed to have no means of conveying to a cell the huge variety of instructions needed to maintain the process of life. But then along came a teenage prodigy from Chicago University: James Watson.

Enter the bratpack

Watson wasn't then (and isn't now) the sort of person to be easily swayed by what Grand Old Men are saying. In *The Double Helix*, Watson's fascinating – and, when it first appeared, rather scandalous – account of the search for the key to life, he states clearly what he thought of those who remained unconvinced by Avery's research: 'Cantankerous fools who unfailingly backed the wrong horses'.

Watson decided that if there was nothing particularly interesting about the chemicals in DNA, then it must be the *arrangement* of those chemicals – that is, the structure of DNA – which held the key to life.

How could he find out more about the structure of DNA? By going to Cambridge University, the world's leading centre for x-ray diffraction, the prime technique for uncovering the structure of chemicals.

Invented by the British physicist William Henry Bragg and his son Lawrence during World War I, this method works by shining x-rays through specimens and studying the pattern of x-rays that emerges on the other side. The pattern, which to the untrained eye looks like a meaningless pattern of spots and lines, reflects the internal layout of atoms in the specimen.

It is an extremely powerful technique, and in 1915 won the Braggs a Nobel prize. Watson decided that he was wasting precious time on the rather run-of-the-mill postdoctoral work he was then doing, and decided to head for Cambridge.

By the time the 23-year-old wunderkind arrived at the Cavendish Laboratory in the autumn of 1951, Lawrence Bragg was the director of the laboratory. Space at the lab was at a premium, and so Watson was given a room to share with another research student – one who, frankly, was getting a bit long in the tooth to be still without a doctorate: Francis Crick.

Watson described his first impressions of Crick thus: 'At that time he was thirty-five, yet almost totally unknown. Although some of his closest colleagues realised the value of his quick, penetrating mind and frequently sought his advice, he was often not appreciated, and most people thought he talked too much.'

In fact, the two got on famously. To his delight, Watson quickly discovered that his roommate shared his view that the secret of life was to be found in the structure of DNA. During their conversations they came to the view that a reasonable first guess for the structure of DNA was the spiral-shaped geometrical figure known to mathematicians as a helix. The special repeating properties of the helix were already being talked about by many of those working on x-ray diffraction of complex molecules. Indeed, work being done on the structure of the alternative key to life – proteins – had suggested that a helical structure, with all sorts of nice chemical properties, was involved. As Crick put it, 'Helices were in the air, and you would have to be either obtuse or very obstinate not to think along helical lines.'

To build up a picture of how this structure could work with DNA and to see where it led, Watson and Crick needed really clear x-ray diffraction images of DNA. Although Cambridge still did pioneering work in the basic technique, Watson discovered that its application to DNA was effectively the sole property of Maurice Wilkins, a physicist working at King's College, London.

Wilkins's laboratory had by far the best x-ray diffraction images of the structure of DNA anywhere. In November 1951, having been in Cambridge just a matter of weeks, Watson decided to find out what Wilkins knew about the structure of DNA. He went down to London to sit in on a talk about the latest x-ray diffraction images presented by Rosalind Franklin, one of Wilkins's team. Characteristically, Watson found himself most interested in what Franklin would have looked like with her glasses off and some decent hair-styling, and thus failed to make any notes. Sure enough, when he returned to Cambridge for a grilling from Crick on the DNA data, Watson could give only a garbled account of what he had heard.

This was to lead to Crick and Watson making fools of themselves. Using Watson's faulty data, they decided that the structure of DNA was essentially three helices wrapped around one another. They mustered their arguments, and then called up the King's team, and invited them to see their brilliant solution to the problem of the structure of DNA.

The London team duly arrived. It soon became apparent that Watson's defective memory had led them to jump to a premature conclusion. By the time the King's team caught the return train back to London, the triple helix idea had died an embarrassing death.

Like two detectives being hauled over the coals for going too far, Watson and Crick were ordered off the 'DNA case' by their superiors, and told to get on with some proper work. And, just like the two hero detectives of fiction, they ignored the warnings, and continued working on the case. By the summer of the following year, 1952, Crick had had an idea about how the DNA molecule might be able to carry out one of the key functions of a cell – 'replication'. This is the process which enables a dividing cell to pass on its protein-making information to its offspring. The pair worked out the details, and this time left the idea to stew for a while.

Their uncharacteristically laid-back approach evaporated immediately when, in mid-December, alarming news arrived from the United States. Linus Pauling of the California Institute of Technology, the world's leading authority on how chemicals combine together, was rumoured to have solved the mystery of the structure of DNA.

If he had the structure, then chances were he also had the key to life, and had worked out how the structure of DNA enabled it to do what the proteins could not. Alternatively – and worse still – he might have discovered that the structure of DNA was truly boring, and that Crick, Watson and the others who thought DNA held the key to life were just wrong.

When Pauling's paper putting forward his theory finally arrived in Cambridge, Crick and Watson breathed a sigh of relief. Pauling was putting forward a three-helix model for DNA, similar to their own failed attempt of a year before. It soon transpired that, incredibly, the world's leading expert had goofed.

Things then began to move very rapidly. Watson, on another trip to see the King's group, was shown new x-ray diffraction images of DNA that provided the best evidence thus far that DNA was a helical structure. It was now a question of how many of these helical chains were wrapped around each other. This time, Watson had made a few notes of what he had heard from the King's group, and he tried to build a model of the DNA that was compatible with their results.

Watson reckoned that since three helices was out, two was the next best bet. After all, didn't biological systems often go in pairs? The trick was to find a way of getting a chemically sensible structure that contained the 'bases' A, T, C and G, held together by a chemical backbone.

After a week or so of fiddling, Watson hit on a structure that seemed to work. It was a bit messy, but after many months of searching for an answer, Watson was very happy. He mentioned his marvellous discovery in a letter to Max Delbruck, a leading researcher in genetics based in California, and then went off to the Cavendish to tell his colleagues about his breakthrough. But no sooner had he begun explaining it than a fellow American – Jerry Donohue, an expert in crystal structures – told him his idea was bunk. Donohue explained that the chemical arrangements Watson had taken from his textbooks to construct his model were mistaken: the textbooks, according to Donohue, were wrong.

Watson knew he had to take Donohue seriously. 'I couldn't kid myself that he did not grasp our problem. During the six months that he occupied a desk in our office, I had never heard him shooting his mouth off on subjects about which he knew nothing.'

So Watson went back to his office, feeling dejected and in no mood for work. He contented himself that dreary afternoon with making some cardboard models of the bases A, T, C and G. He made them accurate enough to tell where chemical 'hydrogen bonding' could occur, as this would help uncover which pairings of A, T, C and G would make a decent double helix.

This was an important clue to unravelling the secret to life. But Crick also believed there was a second, discovered in 1952 by an Austrian biochemist named Edwin Chargaff. He had found that the total amount of A and G in DNA appeared to be the same as the total amount of T and C, and that the amount of A was always the same as the amount of T, while the amount of C was always the same as the amount of G. Despite Crick's interest in Chargaff's discovery, Watson thought of it as being little more than coincidence.

But his attitude soon changed. The next morning, Watson cleared a big space on his desk and started playing with the cardboard models again, to see if there were any combinations that would produce a workable double helix structure. What happened next is one of the greatest moments of discovery in the whole of science. 'When Jerry came in I looked up, saw it was not Francis, and began shifting the bases in and out of various other pairing possibilities. Suddenly I became aware that an adenine-thymine pair held together by two hydrogen bonds was identical in shape to a guanine-cytosine pair held together by at least two hydrogen bonds. All the hydrogen bonds seemed to form naturally; no fudging was required to make the two types of base pairs identical in shape. Quickly I called Jerry over to ask him whether this time he had any objection to my new base pairs. When he said no, my morale skyrocketed.'

Watson could see he had been wrong about Chargaff's work. The pairing of A with T and G with C instantly explained the Austrian's results. Better still, the 'pairing' system seemed to offer a chemical explanation for how DNA could make a perfect copy of itself. The sequence of bases on one strand of the double helix uniquely determines the sequence of bases on the other. If one strand splits off from the other during cell division, it can always reconstruct the strand it left behind by using Chargaff's base-pairing rules.

When Crick arrived, Watson hurriedly told him the news. Crick tried a few other combinations of the cardboard models, but saw that only Watson's structure worked. He also noticed that it strongly suggested that the two helical strands of DNA ran in opposite directions to one another.

However, there was still room for doubt and Watson felt that, given the blunders he had made of late, he ought to exercise some caution this time. Crick, however, had no such doubts. Watson recalls that by lunchtime Crick was in the local pub announcing to all present that he and Watson had found the secret of life. This time, in March 1953, our DNA detectives really had got their man.

Watson and Crick's paper explaining the double helix structure of DNA appeared in the prestigious science journal *Nature* in April 1953. Its uncharacteristically restrained tone reflects the fact that, even at this stage, Watson was fearful of making an idiot of himself. The most exciting implications of the double-helix structure were mentioned almost in passing, right at the end of the paper: 'It has not escaped our notice that the specific pairing we have postulated immediately suggests a possible copying mechanism for the genetic material.' Such coyness would almost certainly be thrown out today. Indeed, given the appearance in the very same issue of *Nature* of x-ray results from the King's team giving convincing experimental support for the claims, Watson's fears seem almost neurotic. In fact, Crick admits that the model put forward in the *Nature* paper did have some remaining problems. Not until the early 1980s was the double-helix structure of DNA finally confirmed beyond all doubt.

But Crick and Watson did not have to wait this long for their reward. In 1962, together with Maurice Wilkins, they won the Nobel prize for physiology and medicine.

How the key to life works

The discovery that DNA is, in fact, a double helix structure with a clever base pairing system showed that DNA could indeed carry at least some information. The pairing mechanism enables a perfect copy of one strand of the helix to be made from the other strand as required. This neatly explains how one cell is able to pass on its protein-making instructions to the others formed when it divides. The DNA strands separate and one strand is passed on to the off-spring, which then uses the pairing rules to reconstruct the strand it left behind.

Having two strands is also an excellent way of protecting the instructions from damage. Ultraviolet light and some chemicals such as benzene have the ability to damage DNA in cells, for example by fusing two bases together. However, if the damage occurs only to one strand, DNA can use the base sequence in the other strand to reconstruct the damaged base sequence. Enzymes exist within cells to snip out defective bases on one strand and insert the correct ones again, using the undamaged strand as a 'crib'.

However, sometimes this mechanism is not enough and some of the instructions are irrevocably changed. The cell is then controlled by faulty data, and this can be fatal: cancer is one possible result.

After the discovery of the structure of DNA, Crick and other 'molecular biologists', as they became known, turned their attention to the obvious next question: just what form do the 'instructions' of DNA take – and how do living cells act upon them? The story of how they reached the right answers is convoluted and confusing. So, rather than recounting the story in strict historical order, let us systematically attack the problem, on the assumption that the researchers in the laboratories have all the data we need to help us. Crick and the others did not have this advantage; what follows serves only to make their breakthroughs all the more impressive.

The mystery to be solved is this: how can the tens of thousands of different proteins found in living things be made from instructions in the form of different sequences of just the four bases A, T, C and G?

Faced with complexity, the usual scientific approach is to throw away as much of the messy detail as possible and develop a theory that explains the broad outline of the problem. One then sees how the theory copes when increasing amounts of the complex detail are put back in again.

So – first question: what are all these thousands of incredibly complex proteins made from? The chemists had a ready answer: chains of chemicals called amino acids. Amazingly, it turns out that all these myriad proteins are made up of different combinations of just twenty different amino acids. We can make about half of these amino acids in our bodies, but over millions of years we have lost the ability to make the others. These we must obtain directly from our diet. (Bacteria, oddly enough, are much better equipped than us, and can still make the full complement of twenty themselves.)

Already our problem of explaining protein synthesis has become far simpler. Instead of having to account for the production of each of the tens of thousands of different proteins, we have only to explain how the four bases in DNA can provide instructions to perform tricks with the twenty different amino acids making up all the proteins. Different combinations of amino acids produce different proteins. This implies that the instructions for making a protein must really be a way of dictating the sequence in which its constituent amino acids are bolted together. The obvious candidate for these combining instructions is the sequence of base pairs on the DNA molecule.

How many base pairs do we need to make an amino acid? One base on its own cannot do the job, as we could then make only four different amino acids – one for each base. What about two bases, like AT or CG? Still not good enough. A bit of thought shows that there are still only sixteen possible arrangements of bases.

Taking bases three at a time, like ATT or GGC, will do the trick, however: there are 64 different permutations of three bases – more than enough to account for all the amino acids we need to manipulate. We even seem to have enough permutations for a bit of 'punctuation' in our code.

A sensible theory of how DNA dictates which proteins are made therefore starts with the idea that a set of just three neighbouring bases on one of the strands of DNA is the basic 'information packet' in the manufacture of a protein by a cell. Proteins are then built up using instructions in the form of sequences of these triple-base units, known as 'codons'.

What a gene really is

Years before the structure of DNA was known, biologists had coined a word for the basic unit of information which controls a cell and which is passed on from generation to generation: the gene. For years, the idea of a gene was rather vague, but now we can define it precisely: it is a sequence of codons, along a strand of DNA, which carries the instructions for (i.e. 'codes for') the manufacture of a protein.

How long is a typical gene sequence for a protein? This depends on the size of the protein, of course, but a typical protein consists of a few hundred amino acids strung together. So, as each amino acid has a three-base-pair representation, the number of base pairs in the average gene sequence is about 1,000.

But just how are these codons used to build proteins? Clues come from looking at the interior of the living cell. The manufacture of proteins takes place outside the nucleus, away from the source of the instructions. So we are looking for something that can act as a messenger, carrying information from the nucleus to the protein-making factories – the ribosomes – sitting outside the cell's nucleus.

Early experiments found no trace of DNA in the ribosomes, apparently ruling out this molecule as the messenger. However, the experiments did show that a chemical closely related to DNA, known as ribose nucleic acid (RNA), appeared to play a role in protein manufacture. Active cells contained more RNA than dormant cells and large concentrations of RNA were found in ribosomes. So is RNA the messenger? Chemical analysis of RNA reveals that it can certainly carry the same amount of information as DNA. Like DNA, it is made up of four bases arranged in sequences. However, there are some differences. The type of RNA found in ribosomes is only single-stranded. In addition, one of the DNA bases – thymine – is replaced in RNA by another compound, uracil. Thus information is stored on RNA in sequences of A, U, C and G.

Even so, one can still envisage RNA using a base-pairing mechanism like that of DNA to make a faithful copy of the DNA's

How a cell builds a protein using DNA

Amino acids derived from food

2 Transfer RNA seeks out certain amino acids

3 tRNA with amino acids attached heads for ribosomes

Nucleus

4 Protein chain emerges

Ribosome, which bolts together amino acids to make protein

DNA master blueprint unstitches to give access to certain genes

1 Messenger RNA, a copy of part of DNA, passes out of nucleus

instructions. And sure enough, RNA turns out to be the messenger molecule which carries information from the nucleus to the protein-making ribosomes.

The process by which RNA takes its message from DNA and transfers it to the ribosomes is, as with so many processes in living creatures, almost miraculously clever. First, a special protein – an enzyme – 'unstitches' the two strands of the DNA 'master instructions' at a certain point. Another enzyme – named transcriptase – then copies ('transcribes') the bases on the exposed stretch of the DNA into a single (and relatively short) strand of RNA. Carrying its copy of instructions from the DNA in the nucleus, this messenger RNA (mRNA for short) then passes out into the cytoplasm and meets up with the ribosomes.

We are now just one step away from explaining from start to finish how DNA controls the life-giving process of protein manufacture. We have to explain only how the amino acids waiting outside the cell's nucleus are finally linked up in the right order to produce the various proteins. (In the jargon, we are talking about building a protein out of short sequences of two or more amino acids – known as peptides – which have themselves been put together in batches called polypeptides).

The amino acids are marshalled together in the right order for processing by the ribosomes by a second form of RNA, called – naturally enough – transfer RNA (tRNA). There are many different types of tRNA molecule, but they all have one thing in common: they can hold on to a particular amino acid at one end and lock on to a specific 'instruction sequence' of three bases at the other. Not just any sequence, of course: there is a relationship between the amino acid the tRNA molecule can latch on to and the set of three bases at its other end.

So, as the ribosome reads off a set of three bases from the mRNA, a suitable tRNA molecule – gripping its amino acid – moves in. The base of the tRNA matches up to the appropriate set of bases on the mRNA, and the amino acid at the other end of the tRNA is left poking up (see diagram on previous page). The ribosome then moves on to read another set of three bases. Another tRNA molecule with another amino acid moves in, and enzymes get to work bolting the first amino acid to the second. A chain of amino acids (a polypeptide) is thus formed.

The process is repeated until the ribosome encounters a sequence of 'punctuation' bases. The ribosome then stops joining any more amino acids on to the chain and the freshly completed polypeptide curls itself up (in ways still not fully understood) into a shape dictated by its composition. The result is a protein which will go on to perform some vital function such as fighting illness or building up a young child's body.

So now, finally, we have the broad picture of how the living cell works. We now know that inside virtually every living (eukaryotic) cell is a nucleus containing objects called *chromosomes*. Human cells have 46 chromosomes, 23 of which come from the father, and 23 from the mother. They consist of special proteins and the double-helix molecule DNA, which carries all the information needed for the cell to function in the form of long sequences of base pairs. These sequences constitute *genes*, and can be passed from one generation to the next aboard the chromosomes carried by the male sperm and the female egg.

The instructions are encoded in genes are converted into messenger RNA, which passes out of the nucleus to the protein-making ribo-

From living cell to genetic code

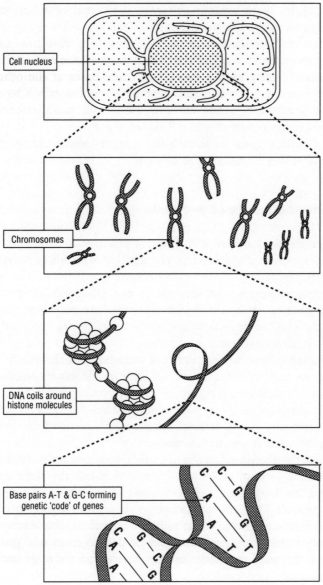

somes. Transfer RNA molecules with various amino acids stuck to them then match up with the mRNA as it is 'read' by the ribosomes, with the sequence of bases on the mRNA dictating which amino acids combine with which. Enzymes then stick the amino acids together, and the result is a protein.

The complete process of protein synthesis we have now outlined constitutes what Crick called the central dogma of molecular biology: DNA makes RNA makes protein. So, change the instructions in the DNA, and you change the capabilities of the cell. Change the capabilities of the cell, and you may have a dramatic effect on the whole organism of which the cell is just a part.

And this fact gives us substantial insight into a major topic of debate: how did humans come to exist?

DNA and the theory of evolution

Most people are familiar with Darwin's theory of evolution, which explains how complex creatures can develop from simpler ones. The key to understanding evolution is recognition of the role of *entirely random* alterations – mutations – in the DNA of living cells. By providing variety in the attributes of living creatures, these mutations enable the famous process of natural selection – the 'survival of the fittest' – to drive evolution onwards.

The vast majority of the random mutations in the DNA do not produce an improvement in the abilities of the host organism. But some do. They enable an organism to survive in its environment, find a mate and reproduce, thus passing on its advantages to its offspring. As the generations pass, natural selection of those with the best gene combination produces improvement.

Talk of the theory of evolution often conjures up mental images of something like a big dinosaur eating a smaller, less fit one. But evolution is taking place here and now. A marvellous illustration of this is provided by the peppered moth, *Biston betularia*. As the result of chance mutations accumulated over the years, there are two varieties of this moth, a dark type and a light type. During the Industrial Revolution, moth fanciers noticed that the lighter type – once far more common than its darker relative – seemed to be disappearing in industrial areas. Out in the country, however, it was doing as well as ever.

The reason was that the pollution in towns was killing lichens on trees, leaving the tree trunks bare and blackened. But mutations had given some moths a light coloration – excellent camouflaging against predators when sitting on lichen, yet worse than useless on the soot-covered trees of towns. The light-coloured moths were easily seen by birds, and were being eaten.

However, its darker cousin was very well suited to the new environment and, unlike the light moth, survived long enough to breed – thus passing on its dark-coloration genes to its progeny. Natural selection therefore boosted the numbers of darker moths in the industrial areas.

The story has a happy ending for the light-coloured moths, however. In the last few years, as a direct result of the clean-air policy now widely adopted in towns, their numbers have been on the increase again.

Exactly the same processes – the transmission of beneficial mutations to our progeny – has enabled humans to become rulers of the planet. We cannot yet control these mutations, however. In the early nineteenth century, French naturalist Jean-Baptiste de Lamarck put forward the idea that by, for example, building up an impressive physique, it was possible to pass it on to our offspring. It's a nice idea, but the work of Crick, Watson and others suggests it is not true. In his book *The Blind Watchmaker*, Richard Dawkins, the Oxford zoologist, uses a nice analogy to explain why. Think of genes in our cells more as a recipe than a blueprint. Then the idea that it is possible to pass on to our offspring traits we acquire during our lifetime is like baking a cake, cutting a slice out and then trying to alter the recipe in the cookbook in some way so that next time we bake a cake, it too will have a slice missing.

That said, over the last year it has shown that, in bacteria at least, there may be more to mutations than pure random changes to genes. In some intriguing experiments (often, and wrongly, considered support for Lamarckism), Professor Barry Hall at Rochester University in America has found evidence that bacteria can somehow dictate which mutations occur within their DNA. When threatened with death by being denied a certain vital nutrient, certain mutant bacteria seem capable of altering their DNA instructions to enable them to make that nutrient. Quite how they do this, no one yet knows.

Putting the key of life to use

Our newly acquired understanding of the workings of the living cell can take us further still. With the key to life, we can use the protein-making abilities of living cells for our own ends.

To do this, we need to build up genes for the protein we want by picking and mixing genes already available on the DNA molecule. This is precisely what is done in the techniques of genetic engineering, in which 'recombinant' DNA containing instructions for the desired protein is built up using special chemicals. We could then plant these tailor-made instructions in a living cell, a bacterium, say, and sit back while it dutifully churns out the required protein.

For decades, the huge chemical plants of multinational companies such as ICI have been manufacturing new drugs of greater or lesser efficacy. Without doubt, the chemistry used inside the production plants is impressive. But, despite all their expertise, the chemical companies have never come close to building a factory that can routinely churn out a huge variety of compounds as complex as proteins. Indeed, it was not until 1990 that the first completely artificial enzyme was produced in the laboratory.

But now, because of recombinant-DNA technology, chemical companies do not need to struggle to imitate the abilities of cells: they can use them directly.

To help to engineer genes, scientists have developed special chemical tools, special bits of DNA called gene probes that home in and pick out just one particular sequence of base pairs along a length of DNA. This is very useful in trying to work out the gene code for a particular protein. A set of proteins, called restriction enzymes, also enables scientists to cut the strands of DNA at a specific point.

To put a new piece of DNA into a cell, one can use a kind of minichromosome called a plasmid. A new bit of DNA is stitched into the plasmid using enzymes, and the combination will then be taken up by a bacterium, and acted upon. When it multiplies, the bacterium will also obligingly pass on the new instructions to all its offspring.

Using such tools, scientists have been able to mass-produce complex proteins existing only in tiny amounts in living creatures, and impossible to make at all using conventional chemistry. The 'protein factories' used depend on the sophistication of the compound to be made. To produce huge quantities of a relatively small protein, for example, the gut bacterium *Escherichia coli* is used. For more complex proteins, animal cells are needed.

All this is not just academic, either. The first products from the genetic 'factories' are already here. The first was human insulin for diabetics. Insulin is a protein hormone, normally produced by the pancreas, which controls the way sugar is used by the body. Diabetics suffer from sugar accumulation which leads to many problems, including blindness.

For years, animals had been the prime source of insulin for diabetics because, although it wasn't the genuine human-derived article, it was relatively easily obtained and worked reasonably well. With the advent of genetic engineering, it has become possible to manufacture artificial human insulin from *E. coli* 'factories' which produce the hormone according to recombinant-DNA instructions.

Recombinant DNA has also helped in more complex situations, where the protein causing the problem could be obtained only from humans. For example, in Britain every year about a hundred children suffering from a lack of growth hormone are born. Without this hormone, they could expect to grow only to a height of a metre or so. Treatment with growth hormone obtained from cadavers seemed to work, but in the mid-1980s fears grew that something was being passed to the patients along with the growth hormone. Some patients were developing an extremely rare and mysterious condition which turned the brain into a spongy mass, resulting in dementia and death. The exact cause remains unknown, but an alternative to the cadaver material had to be found. Fortunately, recombinant-DNA techniques have now led to the development of pure growth hormone produced by laboratory cells, thus completely eliminating the risk of brain disease through the mysterious infection.

Over the coming decades, many more important compounds can be expected to emerge from recombinant-DNA technology. These will include treatments for cancer and AIDS – subjects which we will return to later. However, a major problem with all these treatments is that they are proteins, and, as such, cannot be taken in pill form. Enzymes in the gut will break up the protein, preventing it from being absorbed. Not surprisingly, drug delivery is one of the principal research targets for the pharmaceutical companies.

The key to the truth

One of the most spectacular developments made possible by genetic engineering has been the advent of the DNA fingerprint – arguably the biggest single breakthrough in forensic science this century. It has already led to the arrest and conviction of rapists and murderers. It has also enabled many immigrants to prove their right of abode in a country in the face of scepticism from civil servants.

The technique was invented in 1984 by Alec Jeffries, a young professor at Leicester University. He noticed that human DNA contains long stretches of base pairs repeated many times. The function of these 'stuttered' regions is a mystery, but their length, the number of times they repeat and their exact location is unique for each of us (except for identical twins, who came from the same fertilised egg and thus share the same genes).

Jeffries developed a technique that turned this curious feature of DNA into a vital tool in forensic science. Police can now take a sample of, say, blood or semen from a suspect and send it to a laboratory, where its DNA is extracted. A restriction enzyme is used to split the DNA into fragments of different lengths which are then separated according to their size. Then a gene probe tagged with a tiny source of radioactivity is washed over the fragments. This binds to one specific sequence of bases on each of the fragments. Special film is then used to reveal just where the probe has stuck.

The DNA fingerprint shown in the courtroom consists of a photograph showing a series of dark bands. These indicate where the gene probe has stuck to the fragments. This is compared with a similar DNA fingerprint made from, say, blood taken from the scene of the crime. It too will show dark bands. The more these coincide with those bands on the suspect's DNA fingerprint, the more likely it is that the suspect was the source of the blood or semen at the scene of the crime. For example, if eight bands match the odds against a false identification are 65,000 to one. But if fourteen bands match, the odds are 268 million to one.

The technique was patented in 1987 and proved its worth that same year, playing a central role in the first 'genetic manhunt' in history, set up by British police to trap the man who raped and murdered two fifteen-year-old girls in Leicestershire.

More than five thousand men living in the vicinity of the murders were brought in for biochemical screening, first using conventional techniques, then using DNA fingerprinting. Realising that there was little hope of being cleared by such tests, one man, a 27-year-old bakery worker named Colin Pitchfork, tried to dodge them by getting a friend to give a blood sample in his place. The attempt to deceive was uncovered, however. When blood and DNA tests were performed, Pitchfork was confirmed as the murderer. In fact, the case was a double triumph for DNA fingerprinting. A seventeen-year-old youth who had earlier confessed to the crime was released when his DNA fingerprint failed to match those found at the scene of the crime.

The book of Man

For many molecular biologists – especially James Watson – one of the most exciting future developments is the determination of the entire genetic blueprint – or 'genome' – for a human. Knowing the sequence of base pairs along the DNA would – broadly speaking, anyhow – show how to build a human from scratch.

The enormity of the task facing those who would give us the genetic recipe for a human is breathtaking. Even the lowly bacteria have genomes that are several millions of base pairs long. The genome found in each human cell consists of about three *billion* (3×10^9) base pairs. We get our 'human-ness' from a surprisingly small proportion of these base pairs: 30 per cent of our genome is more or less the same as that of leaf spinach. About 98 per cent of it is the same as that of chimpanzees.

It is the base pairs making up useful genes which are attracting most interest. But we now know that the vast majority of DNA base pairs don't have any clear genetic purpose. As little as two per cent of the base pairs in human DNA actually code for the manufacture of proteins. 'Useful' sequences that do code for proteins are broken up by long stretches of apparently useless base-pair sequences, called introns. The human genome contains far more introns than a bacterium, and some of them are much longer than the useful sequences actually expressed (the so-called exons). The result is that many useful genes are not long interrupted sequences of base pairs: they are 'split'.

Introns are thought to be the result of random garblings – mutations – of the exon sequences serving a useful purpose. Millions of years ago, DNA may have consisted primarily of junk, with mutations within that junk leading to the relatively rapid evolution of proteins. Humans might be much more interesting things than bacteria precisely because they have a much larger store of junk in their genome from which to evolve interesting mutations.

How does mRNA avoid mixing up useless introns with the exons it wants when it transcribes the instructions from the DNA blueprint? Amazingly, over millions of years, mRNA has evolved an extremely precise way of stripping out all the intron rubbish, splicing the useful exons together and taking only those to the ribosomes.

Scientists have cut through these and other complexities and started to unravel the genetic sequences of some simple organisms such as the simple, one-celled bacterium *E. coli* which lives in our gut. The genome of *E. coli* consists of about four million base pairs, written out on a single chromosome containing about 1.4 millimetres of DNA (in comparison, each human cell contains about two metres of DNA).

But progress is now being made on more complex, multicellular organisms. The first such creature likely to have its complete genetic blueprint written down is a lowly soil worm known as *Caenorhabditis elegans*. A fully grown specimen is only a millimetre long and less than a hair's breadth across but, even so, its genome consists of 80,000,000 base pairs, spread over six chromosomes.

Most of the interesting, gene-rich parts of the worm's genome have already been identified or 'mapped'. Now a team of scientists from Britain and America is to home in on these regions and begin to work out the sequence of the base pairs making up these genes. The bulk of the interesting parts of the genome should be established within the next five years.

One of the most important tasks facing the team is that of finding and testing low-cost ways of automatically reading off base-pair sequences. The team hopes to bring the cost down to less than one pound sterling a base pair. This is cheap – but not cheap enough if scientists hope to tackle the ultimate genetic blueprint: the human genome. Even at 50 pence a base pair, sequencing the human genome will cost well over a billion pounds.

International collaboration is the way forward. More than twenty countries have joined up under the auspices of the Human Genome Organisation (HUGO) to work on the blueprint of man. The £2,000-million project – headed by none other than Jim Watson himself – has already identified and decoded about 2,000 of the 50–100,000 genes on the human genome. Watson, in a typically bold gesture, has set a goal of decoding them all by 30 September 2006.

The end result will not be a gripping read: the Book of the Human Genome will consist of thousands and thousands of pages of base-pair sequences, such as AATCGGATTCCC. So why bother? Supporters of the human genome project have described it as the equivalent of Apollo moon landings. Sarcastic sceptics agree – the Apollo moon landings were a waste of time as well, they say. Even so, the major tangible spin-off from knowing the genetic blueprint of man is likely to be in fighting disease.

Most common diseases have some genetic element. Research has shown that diabetes, high blood pressure, coronary heart disease, schizophrenia, some forms of cancer and even psoriasis are to some extent inherited. Already, molecular biology at the level of individual genes has produced some spectacular advances in our understanding of two notorious inherited genetic diseases, Duchenne muscular dystrophy and cystic fibrosis. DMD causes a progressive wasting away of the muscles, while cystic fibrosis results in the production of large amounts of mucus which clogs up the lungs and other parts of the body. The outlook for the tens of thousands diagnosed as having the disease is – at present – bleak.

But ingenious detective work during the 1980s enabled scientists to pick out the defective genes responsible for these diseases from the tens of thousands of genes in the human genome. The immediate benefit is that biochemists have now devised chemical tests to detect the presence of the defective genes even in people unrelated to known 'carriers'. Just a mouthwash can provide enough cells to enable the fault on their DNA instructions to be detected.

But this is just the start. The Central Dogma of molecular biology tells us that a faulty gene means that the disease must also be linked to a faulty protein. Using the information gleaned from the faulty gene, scientists have now found the proteins that hold the key to DMD and cystic fibrosis. Knowing the proteins involved in the diseases helps in the search for drugs.

Most exciting of all is the possibility of actually altering the defective gene responsible for the disease. In 1990, researchers at the US National Institute of Health were given the go-ahead to make the first attempt to repair genes in humans.

The experiments – still underway at the time of writing – are an attempt to treat children suffering from a genetic disease known as adenosine deaminase (ADA) deficiency. Because the children cannot, as a result of this condition, fight any invading organisms, many die in infancy. The experiment involves taking white blood cells from the children and using a special virus to inject the gene for ADA into the cells' genome. Large quantities of the cells are then produced and put back into the children. Animal experiments suggest that the end result will be a greater ability to fight disease. We will soon find out whether it works in humans.

The key to Pandora's box?

Entire books can be, and have been, written on the awesome power unlocked by our understanding of DNA. Its potential to offer cures for long-standing scourges and answers to outstanding mysteries is dazzling. But it can also mask the huge moral questions thrown up by some of the genetic work now underway. Should we hijack the very essence of other creatures for our own ends? There are already plans in America to breed pigs with organs genetically engineered to prevent rejection when transplanted in humans. Is this right? Who should have access to information held on our genomes? Insurance companies could – in principle at least – know from a genome if someone is predisposed to heart disease, and thus impose huge life premiums. By examining a fertilised egg which has undergone just a handful of cell divisions it is possible to determine the sex of the child produced. Already, specialist clinics are getting requests from parents to carry out such tests to ensure they will get a child of the desired sex. Can this be right?

Again, entire books can be, and have been, written on the ethics of genetic experiments. We can only touch on the subject here. Scientists are, of course, excited by their newly found powers. But to portray them – as is often done in the media – as latter-day Frankensteins is to ignore their own sense of moral responsibility. The present approach is to have ethics committees overseeing research proposals. These may not be ideal but, by having scientists and non-scientists sitting side by side, they are probably the best way we have of achieving progress that reflects well on us all.

OUTSTANDING MYSTERIES
The mystery of birth and death

We begin our exploration of the frontiers of research in biology by looking at the underlying mysteries of the two experiences common to us all: birth and death. These are such common events that it is easy for us to overlook their singular features.

Let us start literally at the beginning: how is it possible for a single fertilised egg, smaller than the dot at the end of this sentence, to produce a creature of such astonishing complexity as a human being?

Such a question often presages a dreadfully dull list of phenomena that has to be committed to memory for some exam or other. But a list of events starting with conception and ending with birth explains precisely nothing: it is just an exercise in labelling. It cannot explain how a developing finger cell knows it is to be part of the fingers and not, say, part of a foot.

Such basic questions are now at the forefront of research in biology. A veritable army of scientists is looking for clues to what makes us into what we are, and Nobel prizes are in the offing. But at bottom, they are trying to answer these two questions: how a cell knows where it is, and how it knows what it should be doing.

As we have seen, every living cell starts off with a complete copy of the genetic blueprint, or genome, needed to make the complete organism. Thus, any cell could take up the role of any other cell anywhere else in the body. So we might guess that at least part of the answer to our questions lies in the existence of gene 'switches' that can control which genes out of the tens of thousands available on the genome are actually allowed to operate in particular circumstances.

Second, it seems reasonable to expect that there must exist some form of communication system linking cells up to their neighbours. This would enable cells to act in concert to produce, for example, internal organs of living creatures.

Scientists have been able to test such ideas by carrying out experiments on simple animals such as a conveniently fast-breeding form of fly known as *Drosophila*. The findings apply not just to flies, though – discoveries about their development do have significance for much more complex creatures, including ourselves.

One of the key techniques being used to solve the mystery of development revolves around the study of mutant flies. These are flies which, because of some fault in their genetic make-up, end up having characteristics different from those of the normal fly. By tracking down the cause of the mutation to some error in the behaviour of the cells involved, one can glean important clues to development.

Long before the existence of a genome had been proved, scientists discovered a mutant fly that had an extra pair of wings on part of its body. What seemed to have happened was that cells in the part of its body that were supposed to grow a set of balancers had become confused about where they were, and thought they were in the segment that was supposed to grow wings.

With the discovery of the role of DNA in controlling cell function, scientists were able to think of these weird transformations as the result of mutant versions of the special gene sequences along the fly's DNA. These special genes are known as 'homoeotic' genes, from the Greek word for 'likeness'. They tend to be highly influential: just a single mistake in the sequence of the homoeotic gene can lead to scores of other genes producing, say, an entire limb in the wrong place.

Scientists wanting to solve the mysteries of development are, not surprisingly, concentrating much of their efforts on uncovering how these homoeotic genes exert their influence. One of the most important breakthroughs was made in 1983, when it was discovered that all these genes have a little 'box' of 180 base pairs of DNA in common. This 'homoeobox' contains the instructions for making a short piece of protein that can stick on to a certain part of DNA. The excitement stemmed from the fact that homoeobox genes looked precisely like 'switches' that turn on or off other genes on the DNA. The homoeotic genes code for the influencing protein, and then the homoeobox sticks a bit of protein to the end to make sure the effects are wrought at the right point along the fly's genome. Dozens of these combinations of homoeotic gene and homoeobox – 'homoeobox genes' for short – have now been discovered in *Drosophila* flies.

Research has since revealed the presence of very similar genes in higher animals, including humans. Known as 'Hox' genes, they – like homoeobox genes – are thought to exert their influence on specific sequences of our own genome as we develop inside our mother's womb. These sequences are now being sought. One day, researchers think, it may be possible to tell which Hox genes affect which parts of us.

But there is more to development than gene sequences. Scientists have also uncovered compounds known as 'growth factors' and 'morphogens' (literally, 'shape-givers') which seem to play a crucial role in the differentiation process – that is, in deciding what cells should do what. Their influence comes at a stage shortly after fertilisation when strange and very radical things happen to an egg. After a series of divisions, the egg responds to some invisible signal and turns into a hollow ball, the so-called blastula. After a while, another unseen signal causes the ball to turn in on itself. Then the egg undergoes a crucial event in the development process: the cells it contains are no longer allowed to do whatever they want. They are allotted to three groups: those that will form skin and nervous tissue, those that will form the digestive, breathing and waste disposal system, and those that will hold the lot together, i.e. muscle, bones and blood vessels.

What can be controlling this complex yet, literally, vital process? As long ago as the 1920s, a German zoologist named Hans Spemann made a Nobel prizewinning discovery: the original blastula contains a group of cells – now called Spemann's organiser – which coordinate this three-way specialisation. He also found that this group of cells seems to send out chemical signals, telling surrounding cells what to do.

But until recently the nature of these chemicals remained unclear. Now researchers have identified a candidate for that chemical signal: a peptide called activin. Experiments suggest that it is the level of activin at a particular location that dictates what a cell will become. A group of compounds called retinoids – which includes vitamin A – are also believed to be capable of determining whether cells will become toes, say, or feet. The retinoids may act directly on the cell nucleus, perhaps by triggering the action of the homoeobox genes.

Once the growth factors and morphogens have done their job of telling the cells what to do, the cells themselves produce proteins that ends the specialisation process. Outside influences on the cell are blocked, and all the genes left switched off during the specialisation process are switched off for good.

Thus a cell told to be part of a toe ends up committed to being part of a toe, with its potential for being anything else permanently deactivated – unless, that is, some chemical finds a way of reactivating it. High levels of vitamin A, for example, are well known to cause birth defects; the risk is high enough for some governments to re-commend that pregnant mothers avoid eating liver, a potent source of vitamin A.

But what gets the whole process of development started in the first place? The oft-quoted assertion that genes control everything a cell does turns out not to be entirely true in the development of a living creature. Experiments reveal that the instructions on a creature's genome do not come into operation until several key steps into the development process. Because of this, some scientists are scouring the liquid inside cells – the cytoplasm – for chemicals which control the very first stages of development.

Recent experiments on the African clawed frog *Xenopus laevis* have revealed that a protein called vimentin in the fabric of cells may control where the very first stage of development – the division of the original cell into two halves – takes place. But, as with so much of this research, exactly how it does this is, as yet, unclear.

There are truly more questions than answers in the area of developmental biology, and we have been able to do little more than skim the surface of an exceptionally deep subject. Anyone wanting to do research in a field which is still wide open, with Nobel prizes to be had for decades yet, could do little better than to try to find out precisely why a toe is not a finger.

Making a science of eternal youth

While many scientists are working to find out how we come to be born the way we are, others are looking at the rather more sombre subject of how we come to die.

As we are made up of cells, our demise is intimately linked to the demise of cells themselves. As cells are under the control of the DNA in their nuclei, could it be that there is a genetic reason why we die when we do? Certainly, the ageing of identical twins – that is, individuals sharing essentially the same genome – supports the idea of a genetic link. Such twins tend to die within three years of one another, whereas for ordinary twins (who do not share the same genome) the difference is, on average, about eight years. The fact that different species have different average lifespans also suggests that genes play a role.

Are we, then, 'programmed' to die? The idea of a set of 'death genes' lurking in our genome is an interesting one – if only because it raises the possibility of eventually having such genes excised from our genomes. But there is a problem with such an idea. The basic aim of evolution is to produce beings that are best able to reproduce themselves. Obviously, the longer one can reproduce, the better. Now, if 'death genes' exist, there will be individuals who, by chance genetic mutation, are born without them. But such individuals should, through their increased reproductive lifetimes, eventually dominate the whole population. So why don't they?

One possible explanation lies in the fact that, as we have seen from our discussion of developmental biology, different genes act at different times. Sir Peter Medawar, a distinguished British biochemist, put forward a genetic explanation of death based on this. Throughout history, our genome has been subject to random mutations caused by a host of effects from radiation damage to mistakes by RNA. Some of these mutations have been good for us, but the vast majority have been bad and, most likely, fatal. Now although, as explained earlier, natural selection will tend to weed out such mutations, it will be considerably less successful in weeding out death genes, which tend to become active relatively late in life. In particular, a death gene that comes into action at the end of an individual's reproductive lifetime may stay in the population more or less indefinitely, because the individual with this late-acting gene has time to reproduce, and pass on the gene, before it takes effect.

So the reason we keel over and die may be that our cells eventually come under the control of these late-acting death genes which have accumulated in our genome over millions of years. There is little incentive for the forces of evolution to save us from such genes – all that matters to evolution is that we live long enough to reproduce. The fact is that after we have reproduced, evolution couldn't care less what happens to us.

Statistics show that these late-acting death genes appear to come into play fairly rapidly after our most fertile period is over. Among Westerners, after the age of 30, the death rate doubles every eight years.

This is not to say that the death genes cannot act very early on. Mutations of the information in the human genome are always occurring, bringing death to some from defects they picked up very early on in life. Childhood leukaemias are a case in point.

Death genes are not the only reason we do not live forever. The longer we live, the longer we are exposed to compounds capable of damaging the DNA instructions governing our cells. Many of these – the ultraviolet radiation in sunlight, the cancer-causing agents ('carcinogens') in cigarette smoke – are familiar to us now. But there are also less obvious sources of genetic damage, such as the oxygen in the air we breathe. Highly reactive fragments of oxygen molecules, called free radicals, can cause cancer and lead to death. To that extent, breathing is bad for your health.

Perhaps as a result of the human genome sequencing project now underway, we might, one day, be able to point to some of the death genes that hasten our death, and perhaps eliminate them. By combining human cells with mouse cells, teams from the National Institute of Health in North Carolina and the Kanagawa Cancer Center in Yokohama, Japan, have recently found strong evidence in the laboratory for the existence of ageing genes. The team exploited the fact that some types of cells, particularly cancer cells, are immortal in the sense that they can recreate themselves forever.

In experiments, the two teams combined the 'immortal' mouse cells with normal human cells taken from foetal lung tissue. Most such hybrid cells undergo ageing, but the researchers made the startling discovery that, by removing one of the human chromosomes from the mix, they could make the hybrids immortal, too. This strongly suggests that the chromosome removed carries somewhere on it a gene responsible for ageing.

It seems unlikely that we will ever be able to avoid death entirely. If genes do not get us, outside influences will. But delay ageing? Maybe. A number of researchers have noticed that laboratory mice fed substantially less than they would naturally eat seem to live longer than – even twice as long as – normal. Quite why this is so is not clear. One possibility is that with fewer calories to process, metabolic processes work better and produce fewer damaging free radicals. Maybe eating less simply reduces the amount of poisons to which we expose our organs. It is too early to say that eating less increases your life span. Suffice it to say, however, that some of the researchers involved aren't eating as much as they used to.

Will we ever beat cancer and AIDS?

In recent years, two diseases have emerged as the most feared causes of premature death: cancer and AIDS. This is not a reflection of their prevalence as killers; far more people die of heart disease. The reason that diagnosis of either is viewed with such dread is the fact that some cancers and all AIDS cases remain incurable and that scientists seem to be making slow progress in eliminating them. Fortunately, the outlook is far from hopeless. Major advances in understanding both cancer and AIDS have followed from molecular biology, and the prospect of really effective treatment is fairly bright.

Cancer has been with us for millions of years. Even dinosaurs are known to have suffered from it. Virtually all multicelled organisms can contract it. One in four of us will develop it at some stage. Cancer was given its curious name early in medical history by doctors who noticed that the swellings – tumours – associated with it had an internal structure reminiscent of the shape of a crab, with its central body and spreading arms.

Ironically, it is the success of medical science in fighting disease in general that has led to the current widespread fear about the prevalence of cancer. Up until the middle of the last century, most people simply didn't live long enough to die of cancer, which is predominantly a disease of later life.

Cancer, for all its complexity and different manifestations, can be summed up simply: it is the uncontrolled and useless growth of cells. The cells become revolutionaries, operating under changed orders from some genetic abnormality. Often the result of this uncontrolled growth is a tumour: a mass of multiplying cells.

Tumours are given different names depending on the type of tissue in which they originate. Thus cancers of organ linings, such as skin and lung tumours, are known as carcinomas, while tumours of the connective tissues – muscle and bone – are called sarcomas (an example is Kaposi's sarcoma, which gained notoriety with AIDS patients). Cancers of white blood cells are called leukaemias. Those of the lymphatic system, in which the body's defences against disease are carried, are called lymphomas.

Cancer is intimately linked to the process of cell growth and function. Indeed, one of the motives behind the intense interest in these aspects of cellular behaviour is the need to 'get a handle' on cancer.

And, as with cell development, scientists investigating the causes of cancer are looking closely at the genome for answers.

We now know that cells are surprisingly reluctant revolutionaries: the path to cancer is a long and tortuous one. But the very first step towards cancer involves the genetic blueprint itself. Carcinogens – which can include radiation, chemicals such as benzene or simple errors of replication by the DNA molecule – lead to a garbling of the genetic instructions. Often, the cell's own repair enzymes can fix the damage. But sometimes they cannot. This does not always matter – exposure to a carcinogen is not enough to trigger cancer. The destabilised or 'initiated' cells will lie dormant for years – perhaps forever. The problem is that 'promoting agents', perhaps those involved in promoting growth in healthy cells, can step in to trigger cell division. Even at this stage, cancer can be halted. Growth inhibitors, again perhaps those involved in normal cell behaviour, can stop the cell going out of control.

Thus there are different levels of cancerous behaviour. If damage has been done to relatively few sites on the genome, the cell will continue to behave reasonably normally and the resultant tumour may grow fairly slowly. However, if damage to the genome has been severe, the cell will behave very wildly indeed.

In the early 1980s, scientists in America discovered that some genes can cause catastrophic harm if they become damaged. The discovery came from an interesting observation. Certain viruses implicated in cancer carried a gene that appeared to have a counterpart in the human genome. As the scientific study of cancer is called oncology, the gene was dubbed an 'oncogene'.

When it was sequenced, this oncogene turned out to be very similar to a gene that coded for the production of a growth-factor protein. Here was a clear molecular link between normal growth and cancer. A gene which normally sits on the human genome to control growth is in fact a proto-oncogene which, if damaged, will lead to cancer.

Since this first discovery, scores of oncogenes have been identified. These can lead to cancer in a number of ways. Some oncogenes code for the production of a faulty type of receptor on the cell's outer wall, making the cell act as if it is always receiving the growth factor. Result: uncontrolled growth. Others can cause a cell to respond to the growth factor it makes, again triggering uncontrolled growth.

But of all the oncogenes, one is generating huge interest among cancer researchers: p53. In its healthy form, p53 is a gene, 1,179 base pairs long, which codes for the production of a protein which suppresses cell growth. But just a single base-pair change to p53 ruins its abilities to put the brakes on cell growth – and cancer can be the result. It is now thought that p53 may play a role in at least 60 per cent of all lung, breast and colon cancer cases – the major killers in the West.

The discovery of oncogenes such as p53 is of major importance in fighting cancer. Take, for example, the cancers of later life such as prostate cancer. If these are the result of the activation of oncogenes, it may be that they occur in later life because they involve the activation of a considerable number of oncogenes, and only the passage of time can ensure that enough damage can accumulate. Research suggests that as many as ten oncogenes may have to be activated to trigger these cancers. If this is the case, then genetic 'screening' for the presence of, say, faulty p53 could help alert people years in advance to the risk of contracting these cancers, before all the other oncogenes needed to develop the cancer have been triggered.

Even if the cancer is taking hold of a patient, it may be possible to attack it by killing those cells that make the faulty p53 protein. Another, more subtle technique would be to switch the faulty gene off. One way of doing this would be to make a fragment of DNA with base pairs complementary to those of the gene. In other words, an A base on the gene would be matched by a T base on the fragment, and similarly for the other two bases of DNA. Injected into the body, these fragments of 'antisense' DNA would then bind to the defective gene, shutting it off. Several drug companies, such as Glaxo, are now taking this approach to cancer therapy seriously.

But the understanding of cancer gained by looking at it as a cellular process will do more than just give early warnings. The discovery that cancer involves a number of separate processes means that scientists have a number of separate ways of stopping cancer developing. The traditional treatment of cancer by drugs – chemotherapy – has long been directed by the fact that certain compounds *seem* to work rather than by a deep understanding of *why* they work. The results of this empirical approach can sometimes be paradoxical, with potent carcinogens being used to treat some forms of cancer.

We are now rapidly approaching a time when our understanding of cancers enables us actually to design and build drugs that interfere with key steps in the development of cancer. Many drug companies are now bringing very fast computers to bear on the problem of drug design. These should also help tackle the key problem of toxic side effects, the bane of chemotherapy today.

But it must be said that conventional medicine is already having impressive results in dealing with cancer. About three quarters of children with childhood leukaemias – made notorious by their supposed link to nuclear installations – make a complete recovery. Two thirds of those with breast cancer successfully defeat the disease. Cervical cancer, caught early enough through screening, is almost completely curable. Ironically, lung cancer – almost entirely caused by smoking, and thus avoidable – is one of the least treatable. Fewer than one in ten ever recover.

Certain cancers, such as primary liver cancer, Burkitt's lymphoma and cervical cancer, are now known to be linked to infection by viruses. This opens up the possibility of one day having vaccines against, for example, the papilloma viruses linked to cervical cancer, which kills 2,000 women a year in Britain alone.

The magic bullet

One of the biggest problems of conventional cancer treatment is ensuring that only cells which have become cancerous are attacked and killed by drugs. In a nice irony, scientists have found a way of turning one of the key features of cancer cells to their advantage in the fight against this disease.

Over millions of years of evolution, the body's disease-fighting 'immune system' has developed an incredibly sophisticated battery of defensive cells that secrete proteins, known as 'antibodies', which can home in on foreign invaders and render them harmless. This sounds an ideal way of attacking cancers and indeed, up until the 1970s, scientists tried hard to 'grow' cells which would produce anticancer antibodies to order. Unfortunately, the cells died before they could produce much antibody. But in 1975, Cesar Milstein and Georges Kohler, working together at the Medical Research Council laboratories in Cambridge, had a bright idea. The cells they wanted kept dying after a handful of divisions. Cancer cells, on the other hand, can, effectively, keep dividing forever. So if you cross the disease-fighting cell with a cancerous cell, might you end up with an endless supply of the disease-fighting cells? It's the sort of neat idea that so often fails to work. But this time it did – and won the pair a share in the 1984 Nobel prize for medicine.

Their discovery opened up the way to the mass-production of single-target ('monoclonal') antibody proteins which seek out only those molecules sitting on the surface of cancer cells. At the very least, the monoclonals can be 'tagged' with marker chemicals which can target the cancer cells and thus locate them exactly. But more exciting is the prospect of a magic bullet – a monoclonal tagged with an anticancer drug delivered just to the cancer cell. The theory may be simple but – as so often with living organisms – the practice is not. There are some complex problems to be solved. For example, the immune system of some patients attacks the very antibodies injected to help them. Clinical trials have, however, shown that some patients with certain cancers have responded well.

But perhaps the most exciting strategy now being pursued by cancer scientists is that of *directly* harnessing compounds that exist in the body to fight disease. Not all researchers are convinced that the body's disease-fighting immune system can successfully combat cancer. But others point to the admittedly rare cases of critically ill cancer patients who stage a recovery and eventually rid themselves of the disease completely.

Researchers – notably Steven Rosenberg at the National Cancer Institute in America – have sought out the cancer-fighting compounds that could be behind such curious cases. They are now trying to use them to fight cancers, either by 'training' the compounds to recognise specific cancers, or by boosting their anticancer properties.

Among the compounds now being investigated are interleukin-2 (IL-2), tumour necrosis factor (TNF) and, perhaps most famous of all, interferon. Interferon – produced naturally in the body to fight viruses – came to prominence in the late 1970s when some scientists heralded it as a major breakthrough. Banner headlines proclaiming a miracle cancer cure followed – and disappointment seemed inevitable. Sure enough, interferon is not a miracle cure for all cancers, but in combination with more conventional drugs, it is proving extremely valuable in treating rare ones.

IL-2 and TNF, so-called lymphokines which play key roles in the human immune system, have also had their share of the media limelight. IL-2 shows promise as a means of boosting the cancer-fighting power of the immune system, while TNF has the power to interfere with the blood supply to tumours, in some cases making them wither away completely. The gene sequences for both IL-2 and TNF have been obtained, and genetic engineering techniques are being used to make the most of their natural cancer-fighting abilities. Short term, there are – as ever – problems to be overcome: despite being 'natural', these compounds still have toxic side effects, and are difficult to administer to patients. Long term, however, their highly sophisticated ability to home in only on cancer cells (something chemotherapy cannot do) makes the natural anticancer compounds very appealing.

These extremely brief glimpses of progress in cancer therapy add up to a single message: there will never be a single 'cure for cancer' but the day when this disease ceases to hold so much dread for us is inexorably approaching.

Unravelling the mystery of HIV

Advances in the understanding and treatment of AIDS may soon lead to similarly optimistic conclusions about this most controversial disease of our time. The Acquired Immune Deficiency Syndrome is precisely what it says – a set of symptoms developed in someone whose disease-fighting immune system has lost its ability to fight the myriad potentially lethal infections that assail us every hour of every day of our lives.

This truly dreadful new plague on mankind first came to light in 1981 as the result of detective work by doctors in New York and Los Angeles. They noticed a curious increase in the numbers of previously healthy young men with a rare form of pneumonia called *Pneumocystis carinii*. Others were developing Kaposi's sarcoma, a rare cancer of the connective tissue. The rarity of these diseases stems from the fact that normally they can both be fought off by our immune system. It dawned on the doctors that they might therefore be dealing with something that can wreck the disease-fighting abilities of humans – a deeply disturbing prospect.

Further detective work by scientists at the Centres for Disease Control in Atlanta, Georgia revealed that the number of requests being made for the drug to fight *Pneumocystis* had increased over the previous twelve months, and that all the requests appeared to have been made for homosexual men. By mid-1982 the pattern of AIDS cases across the United States had revealed something else: whatever was responsible for the disease, it was being transmitted

The lethal Aids virus, HIV

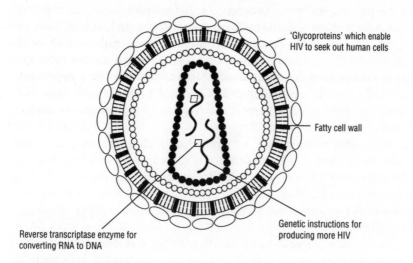

'Glycoproteins' which enable HIV to seek out human cells

Fatty cell wall

Genetic instructions for producing more HIV

Reverse transcriptase enzyme for converting RNA to DNA

by blood exchange and sexual intercourse. Doctors worldwide started to go back over their records, to see if they had ever seen anything like it before. They turned up cases of what is now recognised as AIDS going back to the 1970s.

In 1983, a team under Luc Montagnier at the Institut Pasteur in France isolated from a patient with a closely allied condition the infectious agent now widely believed to be responsible for AIDS. Further studies by teams in the US and the UK confirmed the identification beyond reasonable doubt: the cause of AIDS is a virus, now called the Human Immunodeficiency Virus, or HIV.

Out of Africa?

Testing blood that had been stored in laboratories for years, it was discovered that HIV existed in Zaire during the 1960s. Then, in 1990, scientists at Manchester University's medical school pushed the date back still further, using one of the most important breakthroughs yet made in biotechnology: the DNA amplifier.

Driving through the Californian countryside one night in April

1983, American scientist Kary Mullis had a thought that must surely one day lead to a Nobel prize. Suppose you want to make lots of copies of a small picture. What's the fastest way of doing them? Put the small picture on a photocopier and make a copy. Then put the the copy and the original back on the photocopier, and make another copy. Repeat this process eight times, and you end up with 1,024 copies. Mullis realised he could carry out just such a copying process with special enzymes known as DNA polymerases. As a result, tiny, virtually useless fragments of DNA could be hugely amplified into samples of the original genetic material that were capable of detailed analysis.

Using this polymerase chain reaction (PCR), machines have now been made that in a matter of hours can turn out 100 billion copies of a tiny bit of DNA. The PCR is already proving useful in fields as diverse as cancer diagnosis and forensic science. But one of its most impressive successes to date has been its role in tracing the origins of AIDS.

In 1959, a 25-year-old man died in curious circumstances in a hospital bed in Manchester. A post-mortem revealed that he had died of pneumonia, together with a baffling array of complications and secondary infections. So extraordinary was the case that the pathologist, Dr George Williams, wrote immediately to *The Lancet* describing his findings.

For more than twenty years, Dr Williams thought nothing more of the strange death of the young man. Then, in the early 1980s, news of the spread of AIDS in America started to come in. Dr Williams realised that the young man who died all those years earlier may have been one of the very first victims of HIV.

He approached his colleagues at the medical school and together they scoured the patient records. They eventually found the young man's records. But, better still, they found some tissue taken from the patient which had been stored by the doctors in the 1950s for just such a future re-examination. Could the presence of HIV be found in this tissue? A big problem faced the researchers: HIV is far less prevalent in tissue than in blood specimens. The chances of proving the existence of HIV thus seemed slim.

The polymerase chain reaction proved to be the breakthrough the Manchester scientists needed. By applying it, they were able to detect the telltale signs of HIV in the man's tissue. We now know that the virus can be present in the body for several years or more before eventually undermining the immune system sufficiently to produce AIDS. Thus the implication of the Manchester research is that HIV existed in *the early 1950s*.

This at least puts paid to one theory about the origin of AIDS: that mad scientists at the Pentagon, KGB or some university laboratory genetically engineered the virus in their laboratories and deliberately or accidentally released it on to an unsuspecting world. Back in the early 1950s, Jim Watson and Francis Crick were still trying to work out the shape of DNA; no one then had the knowledge to construct something as sophisticated as HIV.

Clues to the real origin of AIDS come from an analysis of the earliest records of AIDS infection. It now seems likely that HIV came into existence somewhere in Africa. Intriguingly, monkeys in Africa are known to suffer from a condition similar to AIDS. Could it be that HIV is a mutant version of this so-called Simian Immunodeficiency Virus (SIV)? The problem here is that there are now known to be two types of the human AIDS virus, HIV-1 and HIV-2, and they are at least as different from one another as they are from SIV – so much so that they will not trigger AIDS in monkeys (a major handicap in the development and testing of vaccines). If both HIVs had the same origin in monkeys, how have they become so different from one another in the 25 years or so that they have been in existence?

The simplest explanation for the origin of HIV is that it originated in man, from the DNA within our cells. This ties in with the general theory of the origin of all human viruses. At root, viruses are nothing more than a length of genetic material wrapped in a protective protein coat. Such simplicity might suggest that viruses are very ancient. But a little thought shows this cannot be right: viruses are so simple that without a living cell to work on they can do nothing. A virus must invade another cell in order to replicate – the one act evolution cares about. To do this, the virus uses its genetic information to hijack the cell's protein-making facilities. Soon, the cell may be churning out brand-new copies of the original virus which themselves escape, looking for yet more cells to invade.

All this suggests that viruses are in fact rather recent evolutionary developments: rogue sections of DNA or RNA that escaped from a living cell to search for others. Perhaps HIV has been 'on the run' from living cells for centuries, but has only recently undergone the mutations which led to the virulent strain we now see. Its mode of transmission seems important in explaining why we now have an epidemic. To support an epidemic, a critical proportion of the population must have the disease and be capable of passing it on. As it is spread through exchange of bodily fluids, HIV would clearly thrive in situations where sexual activity exceeds a critical level – as happens with the growth of conurbations. Africa has seen just this sort of growth in recent decades.

The scientific war against AIDS

Whatever its origins, scientists now know that HIV is an immensely sophisticated virus. The research into decoding its workings leans heavily on molecular biology – a field which, until the breakthrough of Crick and Watson, simply did not exist. To this extent, we have been fortunate in the timing of HIV's attack on the human race.

It is now known that HIV belongs to a special class of viruses set apart from those responsible for many viral diseases. Inside its protein coat – which measures just 120 billionths of a metre (120×10^{-9} m) across – it carries the information for its replication not in the form of DNA but RNA. To force a cell to do its bidding, HIV must therefore perform a sort of reversal of the Central Dogma of molecular biology, and use RNA to make DNA. For this reason, HIV and similar viruses are known as *retroviruses*.

When HIV invades the body, the disease-fighting immune system responds. What happens next is at present very controversial. The conventional view is that HIV attacks so-called T-cells in the immune system, and breaks into them. The virus then hijacks the T-cells' genome, forcing them to make more copies of the virus. These copies then burst out of the T-cells, killing them and allowing the new copies of virus to go off and attack other T-cells.

Eventually so many T-cells have been attacked and killed that the immune system has been wrecked. Those infected can no longer fight the diseases lurking around us all the time, and succumb to them.

This, then, is the conventional theory of how HIV triggers AIDS. However, during 1991, evidence began to emerge that HIV may act in an altogether more devious way. Detailed studies of molecules on the outside of the virus revealed a small region that bears a striking similarity to a set of molecules on the T-cells it attacks. Furthermore, the T-cells rely on these molecules to help them identify attackers.

This has prompted some researchers, notably Professor Angus Dalgleish and colleagues at St George's Hospital, London, to propose that HIV uses this similarity of molecules to trick T-cells into attacking themselves. Effectively, the immune system commits suicide. If true, this new theory would imply that AIDS is not just another case of disease by viral infection – it is an 'autoimmune' disease, in which the immune system destroys itself.

This new view of AIDS may explain a long-standing puzzle about the disease. If AIDS really is due to the invasion of T-cells by HIV, why is it that when patients with AIDS are examined, only about one in 1,000 of their T-cells actually contains HIV? The new theory explains this neatly: HIV does not have to invade T-cells to wreck them – it gets the T-cells themselves to do its dirty work.

The autoimmune theory has recently gained support as a result of experiments in monkeys and mice which have shown that it is possible to mimic HIV infection by using completely uninfected T-cells. It remains, however, deeply controversial, and the jury is still out.

In the meantime, most scientists are concentrating on finding treatments that look for vulnerable periods in HIV's life cycle. Broadly speaking, the advances are being made on two fronts: finding a way of treating those whose infection has led to their developing AIDS, and finding a vaccine to prevent the virus getting a hold in the first place.

The fact that HIV is a retrovirus is proving a useful chink in its otherwise formidable molecular armour. This is because it is the only cell in the body which has to convert RNA into DNA; it is thus simpler to differentiate it from normal cells. Find a way of interrupting the RNA conversion process, and the biggest blows will be dealt to HIV alone.

Azidothymidine (AZT), the best-known drug in the treatment of those with HIV, is used for exactly that purpose. It blocks the activity of the reverse transcriptase enzyme that HIV needs to make DNA. Unfortunately, AZT – developed years ago as a putative anticancer agent – has side effects, and does not halt the progress of AIDS indefinitely.

Scientists are therefore looking at other ways of attacking HIV, during, for example, the gene transcription stage. Another approach is to stop HIV breaking into helper T-cells. These cells have on their surface a collection of molecules known as the CD4 receptor. This acts like a 'keyhole' for HIV to lock in to. The relevant 'key' on the surface of HIV is another protein, codenamed gp120. This protein is important at two stages in the HIV life cycle. First, it enables the HIV to lock on to a cell it is about to invade. Second, once HIV has hijacked the cell, it orders the construction of gp120, which finds its way up to the cell's membrane, ready to break free as part of a new virus. Thus, invaded cells signal their infected condition by having viral gp120 sitting on their surface.

One way now being looked at of stopping HIV gaining entry is the development of a form of the CD4 receptor molecule that can travel around the blood system, independent of the helper T-cells. This soluble CD4 would then bind to the gp120 on any HIV it encountered, preventing HIV from subsequently binding to the CD4 on healthy cells. One problem with this neat idea is that CD4 survives for only a matter of minutes in the body. To get around this, scientists at Genentech in America combined CD4 with antibodies, boosting its effective life span. Experiments in 1991 showed that primates injected with the modified 'long-life' CD4 remained free of HIV infection after being exposed to the virus.

Instead of simply blocking up HIV's gp120 receptor to prevent the virus entering a cell, scientists at the National Cancer Institute and the National Institute of Allergy and Infectious Diseases in America have been working on the construction of a molecule consisting of part of a cell-killing toxin, produced by a bacterium called *Pseudomonas*, joined chemically to a CD4 molecule. The hybrid seeks out the gp120 on HIV-infected cells, enters the cells and kills them. Experiments in test tubes have shown that the compound does indeed kill off only infected cells. However, there is a lot of work to be done before such techniques can be judged successful in living creatures, and in humans in particular.

The most obvious way of protecting against HIV is, however, to develop a vaccine. The idea is to stimulate the body's own immune system to construct antibodies capable of seeking out and destroying a foreign organism such as HIV. But the trick with an HIV vaccine is a delicate one: the immune system must be given enough exposure to the virus to enable it to build specific antibodies, but not so much that the virus overwhelms the whole immune system.

One obvious approach would be to have a vaccine consisting of killed HIV; however, the risk that some unkilled HIV might slip through makes this approach impractical. Fortunately, it is not necessary to use a whole HIV to raise antibodies against it; the structural protein that makes up HIV's protective coat, with the rest of the virus killed using chemicals, seems to provide enough information for the body's disease-fighting immune system to work with.

Even so, much work on vaccines has involved the use of whole, killed HIV and its monkey-borne equivalent, SIV. Tests with macaque monkeys, using formalin-inactivated SIV, have shown that antibodies do seem to be generated and that these do seem to protect against SIV. The very broad similarity between SIV and HIV, and between macaques and humans, appears to be some cause for optimism about the development of a human vaccine.

However, in 1991, evidence emerged from monkey experiments by Medical Research Council scientists in Britain that threw a spanner in the works. Monkeys given T-cells that had been infected with SIV and then killed were, as expected, found to be protected against SIV. However, monkeys given just inactivated T-cells also turned out to be protected. Although difficult to understand on the conventional model of how SIV (and HIV) works, these results appear to support the autoimmune theory of AIDS that gained ground in 1991.

But even when an effective vaccine is finally identified, there remains a major hurdle: how should a candidate human vaccine be tested? The only real test is to have volunteers injected by a candidate vaccine and then exposed to the full power of HIV, knowing failure of the vaccine meant death. But who would volunteer? One answer is to wait until compounds capable of stopping the spread of HIV within the body become available before conducting human vaccine tests. But can – or should – we wait that long?

None of the developments outlined here will come to fruition fast enough to save the millions of people already infected with HIV. But although AIDS researchers will not put a date on when we will be able to fight HIV effectively, few of them think the day will not come when we have the better of the astonishingly devious HIV. Success against HIV will undoubtedly rank as one of the greatest triumphs of the field which Crick and Watson opened up with their bits of cardboard.

How did life begin?

We now come to the ultimate question for biologists – how they came to have anything to work on at all. In other words, how did life come to exist?

The study of the origin of life is a subject fraught with uncertainties. Despite many years of work and the great ingenuity of those carrying out the work, many of these difficulties have yet to be resolved. There is no generally accepted theory of how life began. Some maintain that we are just missing some simple trick of chemistry to make the whole theory hang together. But many, if not most, biologists think the whole subject so messy, uncertain and complex that it is best avoided altogether.

Let us begin our exploration of this Outstanding Mystery with a few certainties. First, just what are we seeking to explain? From what we have learnt in this chapter about the nature of life, we know the key feature which we must account for: the origin of replicating compounds such as DNA. If we can arrive at a means of creating these out of simpler compounds without a lot of special pleading, then we can consider the problem of life more or less solved. Darwin's theory of evolution, which provides us with the power to get from simple bacteria to ourselves, depends crucially on *occasionally faulty* replication – random mutations of DNA, to be precise – to enable the process of natural selection to work its magic of producing creatures of marvellous complexity from uninspired beginnings.

What clues about the emergence of complex life forms can we glean from the fossil record? This has an important bearing on the mystery, because any viable solution must prove capable of generating the relevant form of life at the right time, using the ingredients then available on Earth. Today's sophisticated eukaryotic cells – that is, ones with a relatively complex internal structure – originated about 1.3 billion years ago. Before then, we find only the simpler procaryotes like bacteria which emerged 3.5 billion years ago. They may have existed even longer ago than that – the problem is that rocks this old are hard to find. The end of the geological record is currently believed to be 3.93 billion years ago, in the tonalitic orthogneiss of Enderby Land, Antarctica, discovered in 1986. The Earth itself was formed about 700 million years before then.

Thus the task before us is to explain how, using the ingredients then available on Earth, to produce something as complex as a bacterium in a billion years or so. Bacteria replicate using compounds like DNA, so at the very least we are asking that our theory explain the origin of something vaguely similar to DNA in a billion years. Can it be done? This could be described as the 'bottom-up' approach to the problem: start with a few basic ingredients and then try to find some way of getting something interesting out of them.

Why not begin by setting up an experiment which mimics conditions on the early Earth, and then see how long it takes for some interesting chemical to emerge?

This is precisely what Stanley Miller, a graduate student at Chicago University, did in 1952. He exposed a sealed flask of boiling water to a combination of gases that his supervisor, the Nobel prizewinning chemist Professor Harold Urey, reckoned would have existed in the early Earth's atmosphere: methane, hydrogen and ammonia. To help boost the chances of getting something interesting out of the mix, Miller passed the rising steam over electrodes which produced an electric spark, a crude simulation of primordial lightning.

After the experiment had been running for a week, something interesting did indeed occur inside the apparatus. The water turned a yellowish brown. However, detailed analysis was disappointing. The most abundant end product was nothing more interesting than tar. Miller tried again, varying the set-up a little. This time the results were much more interesting: among the residues left over from the experiment were amino acids. These are the building blocks of proteins, a key ingredient of living things.

Miller's experiment was instantly hailed as a major breakthrough. Outside the small coterie of biologists working in the field, it is still held to be more or less the answer to the mystery of the origin of life. If so simple an experiment can produce something as important as amino acids, runs the argument, then a bit more experimental sophistication should produce the rest of the ingredients needed to produce a living creature.

Such statements are a triumph of hope over experience. It is the biochemical equivalent of thinking that because we have found a mixture that produces steel, just a bit of tweaking will enable a Porsche to emerge.

For a start, amino acids are interesting – they are what DNA's instructions shuffle around in cells to produce proteins. But amino acids are not DNA, or anything like it. They have no ability to replicate – the central requirement for the chemical basis of life.

Miller's experiment didn't even produce many of the amino acids needed for life. Worse still, many researchers now believe that Miller's basic ingredients were wrong. Using lab experiments and computer simulations of the Earth's early atmosphere, it has been found that ultraviolet radiation from the sun may have destroyed any methane and ammonia in the atmosphere. Take these gases out of Miller's experiment and replace them with the more likely combination of nitrogen, carbon dioxide and hydrogen gases, and the amount of amino acids produced drops radically – in some cases to zero.

But most damning of all is that no uncontrived simulation of primordial conditions has succeeded in producing the sort of compounds needed to build replicating molecules such as RNA and DNA. This is not surprising – water, a chemical virtually certain to have existed on the early Earth, tears apart many complex compounds. To sum up, the Miller experiment did produce a handful of amino acids, but it was far less impressive than many people – scientists included – commonly believe.

Having failed to get very far with the bottom-up approach, let us try going 'top-down'. We know that the living cell today relies on three things: DNA, to carry an accurate copy of protein-making instructions, RNA, to take those instructions to the protein-making parts of the cell, and special enzymatic proteins to get the protein-making process to work. Which came first: DNA, RNA or proteins? If we opt for DNA, we encounter the ultimate chicken-and-egg problem in biology: DNA contains the instructions for building proteins, but it needs certain proteins – enzymes – in the first place to carry out its task of passing on that information. In 1982, Thomas Cech and colleagues at Colorado University made a Nobel prizewinning discovery that promised to cut through the impasse. They discovered that RNA molecules have the ability to rearrange themselves, cutting out various bits and rejoining others, without the need for outside help. It was later discovered that one RNA molecule could also help in the rearrangement of another; bits of RNA could act like enzymes – 'ribozymes', in the jargon. Thus there is little doubt now that RNA preceded DNA, as RNA can perform useful tasks – perhaps even the direction of protein synthesis – without the aid of proteins.

But did RNA exist before proteins as well? There are grave difficulties with this idea. RNA is an extremely complex molecule. It is hard to construct it out of ingredients thought to exist on the early Earth and harder still to get RNA to copy itself – the key requirement. One could still argue that, unlikely as it is, RNA did emerge by chance – after all, a billion years is a long time.

But the arguments against RNA emerging first are certainly strong enough to look at another possibility: that proteins came before both

RNA and DNA. The idea here is that the long road to humans began with the formation of chains of amino acids long enough to do something interesting – that is, to form proteins big enough to act as enzymes to catalyse reactions.

Miller-type experiments, using more realistic simulations of the primordial Earth, can produce tiny amounts of some amino acids. The question now arises of how many amino acids we need to put together to produce an enzyme. Not many at all seems to be the answer – even single amino acids can to some extent act as catalysts. However, the odds against even forming a small number of enzymes, each made up of its own complement of amino acids, is so huge it seems unlikely to have come about at random within the billion years we have allotted ourselves.

But if we turn a blind eye to this problem, the idea of proteins coming first seems to have quite a lot going for it. Professor Robert Shapiro of New York University points out that, while difficult, making RNA out of proteins may not be impossible, given a suitable set of enzymes to help the process along. Modern life is not, however, based on replicating proteins but on RNA and DNA. Thus there must have been some advantages in switching over the role of replication to these nucleic acids. Perhaps it was slower and incapable of producing so much variety. This might explain why, up until the emergence of complex, eukaryotic life forms 1.3 billion years ago, no very exciting life forms emerged. These far more sophisticated life forms became possible once the role of heredity had been passed on from proteins to the nucleic acids used today.

Some origin-of-life researchers believe that current thinking is not radical enough. For example, Dr Graham Cairns-Smith of Glasgow University has put forward a theory based on the idea that crystals may have a crude form of replicating ability. Here 'genes' would consist of particular arrangements of crystal structure, and it is these patterns that would be passed on, generation to generation. Cairns-Smith's ideas centre on the crystalline character of clays but, so far, 'crystal genes' have yet to be demonstrated even in the laboratory. Developments in materials science may yet, however, provide backing for this fascinating possibility.

The field is wide open and may be simply waiting for some clever lateral thinking to cut through the apparent complexity. This is the view of Stanley Miller, who remains convinced that the answer to the mystery of the origin of life will, in the end, be simple. But

Harold Klein, chairman of a recent US National Academy of Sciences committee looking into the current state of research into the origin of life, probably speaks for the majority of biologists: 'The simplest bacterium is so damn complicated from the point of view of a chemist that it is almost impossible to imagine how it happened.'

And yet, against all the apparent odds, it did happen – at least once. Life on Earth has proved remarkably resilient over the aeons, having survived Ice Ages, continental upheaval and, it now appears, calamities triggered by the impact of huge chunks of rock from space. Let us now turn to these matters, and assess just what we can – and cannot – say about the future fate of this planet.

3 A Fragile Shelter in a Hostile Universe

ONE OF THE GREAT ACHIEVEMENTS of post-Renaissance science was the discovery of the law and order that underlies the apparent capriciousness of the natural world. We now recognise violent electric storms as the price you often have to pay for a few days' blistering heat, rather than as the displeasure of some psychotic god. Knowledge – scientific knowledge – is a good antidote for the fear of unknown things.

We now expect science to be able to predict such events, if not now, then in the foreseeable future. The Earth, in short, holds few surprises for us anymore.

But in recent years, an altogether less comforting view of the Earth and our relationship with it has begun to appear. It has been marked by a surge of interest in the ability of the Earth to put up with environmental abuse. Pollution, primarily in the form of carbon dioxide gas from the burning of fossil fuels such as coal and oil, is billowing up into the atmosphere. Acting like the glass panes in a greenhouse, the pollution is trapping more of the sun's heat. This so-called 'greenhouse effect' is slowly but inexorably raising the Earth's average temperature.

So far, the Earth has been sluggish in its reaction to the pollution. But there is growing evidence that, like an overloaded horse shifting under its burden, the environment is undergoing some form of readjustment. That burden is us. So long as nothing happens too quickly, we may be able to cope with the readjustment. But, as we shall see, some scientists are warning that we may be about to be unceremoniously thrown by our ageing mount.

More recent still has come the recognition by scientists that the comforting picture of the Earth as an indomitable shelter in a hostile universe may be illusory. In 1979, American scientists made the astonishing claim that about 65 million years ago the Earth was hit by an asteroid or comet, perhaps even by a shower of comets. The claim was all the more intriguing because the date of the impact coincides with the disappearance of the dinosaurs. Was their hundred-million year reign ended by a cosmic impact?

It now seems the dinosaurs may not have been alone: evidence for mass extinctions of life on Earth, dating back 400 million years and more, has recently been found. These, too, may have been the result of devastating blows from the depths of space. It seems that life on Earth has always had a cosmic Damoclean sword hanging over its head. Intriguingly, the ancients were often terrified by the appearance of a comet; it is no longer hard to see why.

Are we facing devastation, if not by a comet, then amid our own filth and fumes? It is often said that we must look after this planet – it is the only one we've got. But is the Earth unique? Might there not be other planets and other civilisations facing the same challenges? These are the questions we shall investigate. But first we must know a little about what lies beneath our feet, and above our heads.

Journey to the centre of the Earth

For the last 4.5 billion years or so, the Earth has orbited the sun, third of the four inner, rocky planets of the solar system. Just as less dense cream sits on the top of milk, so is the Earth made up of layers of progressively less dense materials 'floating' on one another; at the very centre is the core, containing the highest density material of all.

Can one say anything more than this? Digging a very deep hole would help, but the deepest yet dug, at Zapolarny on the Soviet Kola Peninsula, has reached down only about fourteen kilometres – not even 0.3 per cent of the Earth's radius. It has, however, confirmed something all miners know – the deeper one digs, the hotter it gets. At eleven kilometres down, the Soviet team measured a temperature of 200°C – 100°C hotter than boiling water. Where can this heat come from? For centuries a mystery, we now know the answer: the heat comes from radioactivity. The Earth contains traces of certain elements, principally isotopes of potassium, uranium and thorium, which found their way into the body of the Earth at its formation. They were originally forged in a huge stellar explosion which may have triggered the formation of the solar system itself. Now, however, they gradually decay by radioactivity, giving off heat in the process.

Thus the Earth is a little like a gigantic nuclear reactor. It even produces its own dangerous radioactive pollution at the surface. The uranium decays to give a dense, invisible radioactive gas called radon, which seeps into homes and offices, reaching high enough levels to kill tens of thousands of people through lung cancer every year.

Digging holes is clearly never going to do more than scratch the surface of what happens in the Earth. To find out more, scientists have turned to their most powerful instruments of exploration: pen and paper. They have constructed detailed models of what is going on at depths no human will ever reach.

This 'desktop exploration' had its origins in the early 1800s. At the time, many mathematicians were hard at work developing a general theory of waves: waves in water, waves in the air and, most important for our story, waves in solids. Their academic research turned up an interesting fact: disturb the surface of a solid, and you create two types of waves which go through it at different speeds. So what? Well, an earthquake on the Earth constitutes just such a disturbance. So, adapting the mathematical theory to their own ends, scientists wanting to find out more about the Earth's interior found that it was possible to measure and analyse the behaviour of the two types of waves – now called P- (for Primary) and S- (for Secondary) waves – produced by an earthquake to work out the composition of the Earth. To take an example, theory shows that S-waves cannot travel through liquids. When studying the squiggles on seismograph traces made during some earthquakes, it was found that S-waves failed to appear. From this simple observation, a startling conclusion can be drawn: the outer regions of the Earth's core are not solid but liquid. Some 2,890 km beneath our feet, according to the seismographs, there is a vast subterranean ocean of liquid metal – hardly what one would expect to find deep in the Earth. It is, however, a finding of great importance for the existence of life on the planet, as we shall soon learn.

By 1940, scientists had used the P- and S-wave data to draw up a detailed picture of the Earth's interior. The data were so good that many of the boundaries between each layer throughout the Earth could be specified to the nearest 1,000 metres – and all without so much as lifting a spade.

Since then, the model has been refined still further, using new earthquake data and, latterly, the results of laboratory shock-wave experiments which momentarily generate the collosal pressures found deep in the Earth.

The modern picture of the Earth's interior is as follows: at the very heart of the Earth, 5,150 km beneath where you now sit, lies the inner core, made of a solid, iron-based alloy crushed by gravity to an extraordinarily high density, more than twelve times that of water. A cube of the stuff just 45 cm across would weigh more than a tonne. Pressures inside the inner core reach an unimaginable 3.5 million times atmospheric pressure – about 36 tonnes per square millimetre. Temperatures down there are thought to reach about 4,500°C.

Above the core lies the slightly lower-density outer core. This again is a mix of iron and other elements but, as the seismographs tell us, it is not solid but liquid. How can this be? The explanation is thought to lie in the composition of the outer core: it is made up of an iron compound with a melting point considerably below that of the inner core, making it easier to turn to liquid.

If you still don't believe that there is an ocean of metal beneath your seat, simply hold a magnetised needle up with some thread: it always points in a certain direction. The needle bears witness to the existence of a magnetic field within the Earth, and it is believed that this is intimately related to the liquid nature of the outer core.

When a conducting material moves through a magnetic field, an electric current is produced by the so-called 'dynamo effect'. We are all familiar with it – it enables the lights of a bicycle to be powered by the movement of the wheels. But once the electric current is created, it too produces a magnetic field. It is thus possible to boost a small magnetic field up into a relatively large one.

Such an effect is thought to explain the origin of the Earth's magnetic field. Working from the presumption that a small magnetic field existed in the Earth when it was formed, the theory then calls on heat to provide the motion needed for the dynamo effect to work. The motion would come from convection inside the liquid metal of the outer core producing huge loops of material and triggering a 'geodynamo effect' that generates the Earth's magnetic field.

The basic idea of a geodynamo producing the magnetic field was established in the 1950s, and is thought to be correct in broad detail. However, there remain some puzzles that have yet to be satisfactorily explained. The most intriguing stems from a curious discovery made in 1906 during studies of ancient lava taken from the Massif Central in France. Molten rock itself has no magnetic field. But as it cools, it feels the magnetic influence of its surroundings. Cooling rock thus carries, sealed within it, a record of the direction of the magnetic field of the Earth at the time the rock cooled off.

The lava from the Massif Central had a very strange story to tell. Laboratory measurements showed that when the lava cooled, the Earth's magnetic field was reversed – in other words, today's north magnetic pole was in the southern hemisphere. It was then found that similar magnetic 'flips' have occurred many times during the Earth's history. The most recent occurred about 700,000 years ago. They do not occur overnight – the last one took 10,000 years or so to complete.

The effect of such a magnetic flip on life is uncertain, but is unlikely to be beneficial. The Earth's magnetic field creates a kind of protective force field around the planet – the so-called magnetosphere – which shields us from fast-moving particles spewed out by the sun. Some of these particles manage to get through, however, and smash into the Earth's atmosphere near the north and south poles. The result is the beautiful northern and southern lights – the aurorae borealis and australis.

During a reversal, the protective magnetosphere would shrink and perhaps disappear altogether, allowing these solar particles to reach the Earth's surface. Essentially a form of radiation, the particles might well trigger cancers and gross mutations in living creatures.

Anyone claiming to understand the origin of the Earth's magnetism has now to explain these complete reversals, which occur every 300,000 to a million years, and sometimes in as little as every 40,000 years. Explanations of this bizarre (but, as we shall see, scientifically valuable) phenomenon abound. My own favourite theory pins the blame on so-called 'chaotic' phenomena deep within the Earth. Even relatively simple mechanical devices, such as a pendulum suspended from a moving pivot, can produce weird behaviour, even though the laws governing it are well understood. There is good evidence to suggest that the geodynamo is controlled by processes that may similarly become wild and chaotic if disturbed. The outer core may thus occasionally succumb to these intrinsic bursts of wildness, producing a complete, if temporary, reversal in the Earth's magnetic field.

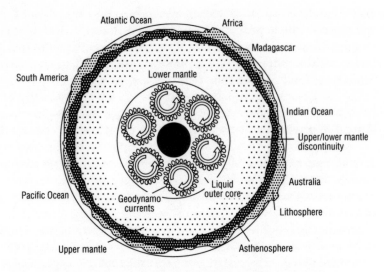

What seismology has uncovered about the Earth's structure
(Not to scale; based on work by Dr John Maxwell, University of Texas)

The giant eggshell

Continuing our journey upward from the core we find, floating on top of the seething outer core, the layer known as the mantle. This consists primarily of relatively light, rocky material at a high enough temperature to allow huge convective currents to circulate, albeit extremely slowly. The uppermost layers of the mantle, however, are relatively cool and form a relatively rigid layer of rock, called the lithosphere, around the Earth. The familiar crust, varying in thickness from 40 km on the continents to as little as five kilometres beneath the oceans, forms the topmost layer of the lithosphere.

The lithosphere is not one smooth, continuous covering of rock, however: it forms a huge broken shell surrounding the Earth, with individual 'fragments', thousands of kilometres across, known as tectonic plates. The origin of these fragments is uncertain; in 1989 Professor Donald Lowe at Stanford University put forward the intriguing suggestion that they were originally formed by showers of huge meteorites smashing into the Earth early in its history.

Despite their huge size, the plates are far from fixed. They form the topmost parts of the convective loops generated deep below, and move slowly about the Earth's surface, carrying entire continents on their backs. The appearance of the Earth is thus in a constant state of change. Continents drift across its face over hundreds of millions of years, occasionally colliding and forming mountain ranges such as the Himalayas, and occasionally breaking up, creating vast new oceans.

Alfred Wegener and his wandering continents

If you find the idea of India moving around and smashing into Asia to form the Himalayas hard to take, you would have been in prestigious company fifty years or so ago. The explanation of continental drift was given a very hard time in its early days, and its slow acceptance is an object lesson for anyone who thinks that science is motivated solely by the search for truth.

For decades its originator earned little but disdain from the geological establishment. He was a German scientist named Alfred Wegener, born in 1880. Wegener was by training and profession not a geologist at all, but a meteorologist – a fact that may well account for much of the initial ill will his ideas attracted.

The starting point for his theory is a simple observation that many children make when studying a map of the world: the east coast of South America and the west coast of Africa seem to fit together like two pieces of a jigsaw. If they were once close together, they would have formed a truly vast 'supercontinent'. In 1910, Wegener put forward just this proposal. He declared that today's continents were once all part of a supercontinent which he called Pangaea, and amassed a variety of geological data to support this idea. Some of the most intriguing were the geological 'folds', now separated by thousands of miles of ocean, that seem once to have been part of one giant system.

But there was a major gap in Wegener's theory: what was propelling the continents about the Earth's surface? He could come up with nothing that seemed capable of breaking up the supercontinent and propelling the fragments about. Most geologists looked on the absence of a decent explanation for continental drift as the only excuse they needed for ignoring the whole of this 'jumped-up weatherman's theory. In 1929, a year before Wegener's death, the British geophysicist Arthur Holmes proposed the convective loop idea now held to be the solution to the propulsive power problem. But it was to take the emergence of a new generation of geologists twenty years later to put Wegener's ideas back on the map.

During the 1950s, the British scientist Keith Runcorn and his colleagues were studying the past history of the Earth's magnetic field, using the magnetic alignments of once molten rocks, when they made a curious discovery. The alignments of rocks in Europe seemed to show that what we now call the north magnetic pole had once been close to Hawaii, and had wandered to its present position via Japan. Runcorn realised that there was another way of looking at this discovery, however: maybe the magnetic pole had stayed just where it is and Europe had moved relative to Hawaii and Japan.

By repeating the measurements around the world, Runcorn found that the magnetic poles seemed to have followed other paths around the Earth's surface. Of course, if the poles had really moved, this was nonsense – they cannot be in two places at the same time. But the findings made perfect sense if the land masses were all changing their relative positions. Continental drift was no longer looking so silly.

Then more evidence started to emerge. Detailed studies of the ocean floor undertaken after World War II uncovered evidence for huge undersea geological features. A towering ocean ridge – Ascension Island – was discovered running down the middle of the Atlantic, and the Azores in the Atlantic turned out to be the peaks of vast mountains that reared up thousands of metres from the seabed. Plunging oceanic canyons were also found, the deepest being the Marianas Trench off the island of Guam in the Pacific, whose floor lies almost eleven kilometres below the waves. (If Mount Everest were put there, its summit would still be more than two kilometres below the surface.)

The American geologist Henry Hess saw that both the towering undersea mountains and abyssal depths could be part of a global geological system. He put forward a detailed picture of what happens to the vast rocky plates taking part in continental drift. Hess proposed that the oceanic mountain ranges were regions where fresh plate material wells up from the mantle at the top of the huge convective loops. Obviously, space has to be made for all this new material. So Hess declared that the abyssal trenches were regions where, to compensate, old plate material sinks back down into the mantle.

Hess's ideas were immediately seen as an elegant solution to a long-standing mystery. Geologists had failed to find beneath the sea any rock older than about 230 million years. Yet the Earth was known to be far older. What was happening to the ancient rock beneath the sea? Hess's theory explained it beautifully: no subsea plate material survived longer than 230 million years before sinking back down into the mantle. His theory went further, however: the subsea ridges, being places where new plate material emerged, should contain only relatively new rock, while the trenches should contain much older rock on its way back down.

By the early 1960s, proof that this was also true finally emerged. Studies by American scientists of rock on the floor of the eastern Pacific revealed a curious stripelike pattern in its magnetic properties, running parallel to the ocean ridges. In 1963, Drummond Matthews and his student Frederick Vine at Cambridge University came up with the explanation: it was the French Massif phenomenon again. Recall that hot, new rock emerging from the mantle at the ocean ridges acquires magnetic properties dictated by the direction of the Earth's magnetic field at the time the rock emerges. When it cools, information about the magnetic field at the time of formation is 'sealed in' the solid rock.

Matthews and Drummond claimed that the stripes of contrasting magnetic properties seen in the subsea rocks were the result of the curious 'flips' that the Earth's magnetic field undergoes from time to time. But most important of all, the width of the alternating bands showed that the sea floor was constantly being supplied with new material: later studies showed that the band widths tied in very nicely with the length of time the various 'flips' were known to have been in force.

One might think that Wegener's ideas would have now been widely accepted, but no. Many geologists thought of the sea-floor magnetic studies as relevant only to marine geology, and took little interest in the new findings. It took some years more before the global significance of the new work was fully appreciated.

When continents clash

The decades of neglect have now given way to a realisation that continental drift can answer puzzles in fields far removed from geology. Why, for example, is there so little similarity between animal populations in South America and Africa, compared to those in

How continental drift has changed the face of the Earth

180 million years ago:
Pangaea super-continent breaking up

120 million years ago:
Tethys seaway has split Pangaea from
east to west; northern region now
formed is Laurasia; southern regions
named Gondwana

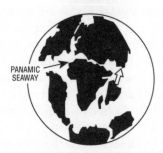

80 million years ago:
Panamic seaway still separates the
Americas; Tethys now being shrunk to
form Mediterranean. India heads north
for Asia

Present day:
Americas have joined; India has
smashed into Asia, forming the
Himalayas in the impact

(Based on work by Dr John Maxwell, University of Texas)

North America and Europe? Continental drift provides a ready expla-nation. About 200 million years ago, the south Atlantic ocean began to open up, splitting the Americas and Africa from one another and dividing animal populations. This happened well before North America split from Europe – so there has been more time for vari-ations to appear.

The past climate of continents has also been greatly influenced by continental drift as well. The most striking example of this is Antarc-tica. Now a desolate snow-covered wasteland, Antarctica contains one of the world's largest coalfields. It must therefore once have been warm enough to support tropical forests. Continental drift has pushed Antarctica to its present position from the much warmer climes it occupied hundreds of millions of years ago.

But the effects of continental drift and 'plate tectonics' can literally be felt today. Although it may take millions of years for continents to drift apart and crash into one another, the process is going on continuously – sometimes with disastrous consequences. When two huge plates grind against one another, huge tension is set up within them. Suddenly the rock can give way and tear, releasing huge amounts of energy, chiefly in the form of violent motion. The result is an earthquake.

A map of where earthquakes occur most frequently reveals a con-centration along the edges of the huge plates, whose slow, inexorable grinding has caused over a million deaths so far this century. Earth-quakes have been studied scientifically for hundreds of years. In AD 132, the Chinese philosopher Chang Heng devised a simple instru-ment that could detect the direction of the first shocks from a quake. By the middle of the nineteenth century, seismometers capable of detecting distant earthquakes had been produced. Today's seis-mometers are sensitive enough to detect the noise of sonic booms from fighter aircraft.

By timing the arrival of P- and S- (and other, more complex) types of waves through the Earth, a network of seismometers provides a wealth of detail about an earthquake. The most well-known piece of information is the so-called Richter magnitude of the quake. It was introduced by the eponymous American seismologist in the 1930s as a means of putting all earthquakes on a similar footing, enabling comparisons of their energy to be made. Two points are worth making about Richter magnitudes, which are quoted so often in news reports. First, contrary to the impression often given, the Richter scale does not run from 1 to 10 – it is open-ended. The feeblest earth tremors ('microseisms') recorded actually have negative Richter magnitudes, while the most powerful quake ever recorded – in Chile on 22 May 1960 – had a Richter magnitude of 8.3.

The second point is that the energy released does not have a simple relationship to the Richter magnitude; a quake measuring 8 on the scale isn't just twice as violent as one measuring 4, but a million times more violent, equivalent in terms of energy released to the detonation of sixteen million tonnes of TNT – a couple of hydrogen bombs worth.

Actually, despite the preoccupation of the media with them, Richter magnitudes are a poor measure of the disastrous effect of an earthquake. What really matters is the density of population near the fault, and what steps they have taken to protect themselves. In February 1960, a comparatively feeble earthquake of Richter magnitude 5.8 struck Agadir in Morocco; it killed about 12,000 people. In October 1989, the huge tension building up between the gigantic Americas plate and the abutting Pacific plate proved too much for the famous San Andreas fault on the Californian coast. It ruptured about six kilometres beneath the township of Loma Prieta. Yet the resulting 7.1-Richter earthquake – almost 100 times more powerful than that which struck Agadir – killed only 69. The reason is that after the Great San Francisco earthquake of 1906, in which 700 died, California was prepared for the worst and buildings had been specially built to withstand earth tremors.

Apart from giving Richter magnitudes, the seismometers enable the depth and location of the centre of the earthquake – the so-called focus – to be found. This is the point at which the tearing of the plates began. From here the tear extends outward, travelling for many seconds, causing tremors all the time. The point on the surface directly above the focus is the epicentre, and is usually named after the nearest township.

The threat posed by earthquakes to major conurbations, particularly in vulnerable California, China and Japan, has led to research into ways of predicting when an earthquake will strike. An intensive study of the geology of the San Andreas fault system by the US Geological Survey led to the prediction – a year before the quake struck – that the Loma Prieta section of the fault was the most likely part of the system north of Los Angeles to produce a magnitude 6.5–7 earthquake.

Although scientifically impressive, such predictions are of little practical use. What is really needed are details of the exact path of destruction a month, say, before the quake strikes. A whole battery of techniques is now being investigated in the search for an accurate way of predicting quakes. They range from measurements of the amount of radon gas seeping out of rock about to fail to the detection of curious bursts of very low-frequency radio waves that preceded the Loma Prieta quake.

The most promising technique centres on the detection of 'fore-shocks' – the faint creakings of rock just before it completely fails. The Chinese used foreshock methods to predict a quake in Manchuria in 1975, although some sceptics say they were just lucky. A vast area was evacuated in time to save thousands of lives. But this is not the triumph of science it might seem. For weeks before the quake struck, the authorities had made other predictions, all inaccurate, and had organised several evacuations. Such false alarms could perhaps be tolerated in a docile, agrarian society. An industrialised nation would find them more than just inconvenient – they could prove disastrous. Many of those killed in the Loma Prieta quake were in cars travelling on a multitiered flyover. If a quake struck sooner than predicted by scientists, far more could be killed on the roads while trying to escape than if they had stayed in town in San Francisco's specially designed skyscrapers.

Although progress is being made, a suitably reliable means of predicting when and where an earthquake will strike is, most seismologists think, decades away at least. In the meantime, the best that can be done is to put a probability on a quake of a certain magnitude occurring in a certain area within, say, the next ten years. For example, Japanese seismologist Professor Tsuneji Rikitake at Nihon University believes that there is a 40 per cent probability of a magnitude 8 earthquake striking the densely populated Tokai region between Tokyo and Nagoya by the end of the century. Tokyo itself is likely to be hit directly by a magnitude 7 earthquake before 2010.

The effect of such an event is likely to be appalling. In 1923, when Tokyo was last rocked by a major earthquake, the death toll was more than 100,000. Even with improvements in building design made in the intervening years, huge numbers of casualties can be expected from another major quake in this densely populated area. The vast, viscid loops of the Earth's mantle will have claimed yet more victims.

The invisible shield

Let us now look above our heads, to the atmosphere. Despite being just a collection of invisible gases weighing just one-millionth of the total mass of the Earth, even barely measurable changes in the behaviour of our atmosphere can have devastating consequences for life. Fortunately, we know far more about the gases over our heads than about the rocks beneath our feet. Unlike the body of the Earth, the atmosphere has been explored from top to bottom, first by balloons and more recently by spacecraft.

Yet there remain gaping holes in our knowledge and there is growing concern that filling those holes is not merely a matter of putting academic icing on the cake of understanding. Indeed, some scientists feel that unless these gaps are filled in rapidly, we may end up with a global disaster.

On the face of it, the atmosphere seems remarkably simple: almost all of it – 99.04 per cent – is made up of just two gases: nitrogen, a boringly inert gas, and oxygen, a highly reactive (indeed cancer-causing, in some circumstances) gas, in a mix of about four parts nitrogen to one of oxygen. The rest is made up of traces of other gases, predominantly argon, another inert chemical, and variable amounts of water vapour. But although some of these latter gases are present in tiny quantities – a few parts in 10,000 of carbon dioxide, for example – they are of literally vital importance.

This is because some of them have the property of letting through the short-wave radiation that comes from the sun and stopping the subsequent long-wave radiation bounced off the Earth's surface from escaping back into space. Because this ability is shared by the glass in greenhouses, those gases which have this heat-trapping ability are known as 'greenhouse gases'. The most important such gas is water vapour, followed by carbon dioxide, methane and oxides of nitrogen.

Without the insulating effect of these trace gases, the Earth would be a radically different planet, suffering vast swings in temperature such as those experienced on the moon. During the day, temperatures would soar to 110°C, and at night plunge to −170°C. The atmosphere smooths out the extremes, keeping the Earth at a more congenial average of 15°C.

But in recent years, a problem has started to become apparent: much man-made pollution also has a greenhouse effect. Are we therefore pushing up the average temperature of the Earth ourselves? Could we trigger unpredictable changes in the climate as a result?

The perils of a hotter world

What we know for certain is that in 1800, before the Industrial Revolution really got underway, atmospheric carbon dioxide made up about 0.028 per cent of the atmosphere. Now it makes up 0.035 per cent – a 25 per cent increase. This appears to be principally the result of burning of fossil fuels such as coal and oil.

The levels of other greenhouse gases have also risen – that of methane, produced primarily by digestion in cattle and insects and farming practices, has doubled in the last 200 years.

But has the Earth's temperature gone up concomitantly? This is one of the most vexed questions in contemporary science. It seems, however, from measurements of the temperature in and over the sea dating back to the 1870s, that the Earth's average temperature has gone up by half a degree or so over the last century. So what's all the fuss about?

The Earth's climate is driven by a combination of the rotation of the Earth and the temperature difference between the equator and the poles. Pollution has the ability to affect the relative importance of these two effects. Worse still, we now know that a small change in temperature does not necessarily lead to a small change in the climate. The atmosphere is, to this extent, potentially subject to chaos – a term we shall consider in much more detail shortly. Thus a small change in temperature may produce a larger jolt to the climate. The climate, in turn, has a profound effect on agriculture, to name but one aspect of life on Earth. Agriculture, in turn, affects the economies and ultimately even the political stability of countries. The historical precedents are not encouraging; it was economic upheaval that enabled the Nazis to come to power. We interfere with the climate at our peril.

Hence the interest in finding out if the Earth's temperature really is rising and, if so, why and how fast. It will be perhaps a decade before we have clear evidence that the Earth's temperature is going up in reaction to man-made pollution. Relatively precise data from orbiting satellites over this length of time is needed to settle the issue.

In addition, even the world's fastest supercomputers cannot yet predict with certainty what a hotter world would be like, and who the winners and losers would be. Making a computer model of the entire Earth's climate is no easy matter, requiring vast computer power and a detailed understanding of all the influences at work. But the majority of climate scientists are already agreed on one thing: we cannot afford to sit around waiting for definitive answers.

In 1989, the United Nations set up the Intergovernmental Panel on Climate Change to investigate the reality and impact of global warming. The consensus of hundreds of scientists is that unless action is taken, the Earth will warm about 0.3°C every ten years. What action is necessary to prevent this? A switch away from fossil fuels seems the obvious answer, but a switch to what? Many people would be reluctant to see an expansion of nuclear power to make up the energy gap. Quite apart from incidents like the Chernobyl reactor explosion of 1986, there are many who worry that even if a nuclear reactor is 100 per cent safe in operation, it still produces waste of a most noxious kind. Renewable forms of energy can provide only a fraction of the energy produced by fossil fuel stations today, and they bring their own problems – wind turbines, for example, are unsightly and create a lot of noise and turbulence downwind.

The long-term answer is far from clear. Maybe efforts by scientists to tame nuclear fusion – the power source of the sun – may succeed early in the next century. The idea here is to build a machine that can force nuclei of light, hydrogenlike materials, to merge with one another, releasing huge amounts of energy that can be converted into heat suitable for generating electricity. Nuclear fusion offers a virtually limitless supply of energy from fuel extracted from seawater. But again, there is a problem of dealing with (relatively short-term) radioactive waste to be solved. Some nuclear scientists even say privately that continuous, controlled nuclear fusion may never be possible on Earth.

Simply conserving – by, for example, insulating homes – energy would be a start. That, at least, would buy us time to come up with some solutions before we cause the climate to do something dramatic.

The hole in the shield

The concern over global warming centres on processes taking place in the very lowest layer of the atmosphere, the so-called troposphere, which extends up to about twelve kilometres. Above it comes the stratosphere, and here we encounter another trace gas about which there is much current concern: ozone.

Ozone is a chemical relative of oxygen, but with rather nastier properties. Even in concentrations of a few parts per ten million of air, ozone has been found to have health effects such as headaches and lung irritation. Such concentrations are regularly reached at ground level in major towns during the summer, where intense sunlight breaks up compounds in car exhaust fumes to create ozone.

But paradoxically the ozone created high up in the stratosphere plays a vital role in protecting life. Along with heat radiation, the sun bathes us in shorter wavelength, and thus more penetrating, ultraviolet radiation. It is this, through a reaction with a pigment in our skin called melanin, which gives us suntans. But too much exposure to the most penetrating UV radiation can also produce skin cancer, through the damage it wreaks on the genetic material inside skin cells.

Stratospheric ozone mops up ultraviolet light and, fortunately, the stratosphere contains enough – about one part per 100 million or so – to eliminate most of the danger. But now hard evidence has emerged that something is eating away at the protective ozone layer.

Since 1957, scientists at the British Antarctic Survey have been carrying out measurements of the amount of ozone in the air over the South Pole. Many would see such research as just the sort of arcana that scientists should not be indulging in at the taxpayers' expense. (Indeed, there were occasional moves by the UK government to end it.) But in 1985 Dr Joe Farman and his colleagues at the BAS made a discovery that was to become all too well known to those same taxpayers. They had found a gaping hole in the protective ozone layer over Antarctica.

The hole turned out not to be permanent but seasonal, and variable in extent. But the hole has now been seen to open wide enough on occasions to expose parts of Australia and New Zealand to much higher levels of cancer-causing UV radiation. Levels of ozone in the atmosphere over Melbourne were found to drop by twelve per cent. This is sufficient to increase the number of skin cancer cases in the area – already among the highest in the world – by more than a third. There are other possible effects. Cases of eye cataracts, a common form of blindness, may increase and the growth of certain crops may seriously deteriorate.

Urgent research has since been carried out to see if a similar hole exists over the north pole. Given the much higher population density in the northern hemisphere, such a hole would have even more serious consequences. So far no actual hole has been found, although airborne surveys carried out in February 1989 suggest that at least 50 per cent of the stratospheric ozone over the Arctic had been destroyed during the winter. Some scientists think it is only a matter of time before a genuine 'hole' is discovered.

So what is the mysterious substance eating away at the ozone layer? It is now thought to be principally a family of man-made gases called chlorofluorocarbons (CFCs). Invented by an American chemist at General Motors in 1930, CFCs were once extensively used in the refrigerator and aerosol-making industries because of their useful combination of properties, including non-flammability and non-toxicity. The problem is that they break down in the atmosphere and release chlorine. Chlorine molecules are potent destroyers of ozone: just one of them can knock out 100,000 molecules of ozone. With CFCs we have a low-concentration gas under attack by an *incredibly* low-concentration compound: CFC levels are usually measured in parts per thousand million of air.

Fortunately, scientists have presented policy-makers with sufficiently worrying evidence to provoke action. International moves to phase out the most destructive CFCs, and replace them with more (but not entirely) ozone-friendly alternatives are underway. But whether or not we have taken action in time to prevent a huge increase in skin cancers and the other effects of higher UV levels remains to be seen. CFCs are slow to reach up into the ozone layer, and once there they stay put for more than 100 years. A complete ban today will not have any effects until well into the next century.

Death from above

Above the stratosphere and its ozone layer, the atmosphere rapidly thins out into the vacuum of space. Where the atmosphere stops and space begins is more a matter of definition than science; it is usually taken to be 100 km above the Earth's surface.

But even at these heights, the atmosphere can exert its protective influence. Subatomic particles spat out by the sun smash into the thin veil of air molecules and cause a cascade of coloured lights which we see as the aurorae (or the northern and southern lights). Meteors pouring in from deep space become heated by friction with the molecules and disintegrate, producing 'shooting stars'.

Fortunately, the vast majority of meteors never survive to cause much damage on the Earth below. But sometimes they do. Giant geological features like the 1.2 km-wide Meteor Crater in Arizona testify to the fact that sometimes huge chunks of cosmic debris have survived the journey through the atmosphere to strike the Earth. And in 1979, a team of American scientists found disturbing evidence that some of these chunks have been big enough to annihilate entire species from the Earth forever.

The story of the discovery of this evidence is every bit as fascinating as the story of Crick and Watson's work on the structure of DNA. And its conclusion is, if anything, even more startling – for it shakes our belief in our Earth as a safe haven in a violent cosmos.

In 1977, Walter Alvarez, the geologist son of the American Nobel prizewinning physicist Luis Alvarez, arrived to take up residence at his father's university, Berkeley in California. On his arrival, Walter gave his father a gift: a small piece of rock that had been cut from a formation near Gubbio, Italy. Polished to show off its layered structure, Walter believed that the rock had some strange story to tell about events on the Earth when dinosaurs reigned supreme, about 65 million years ago.

The rock had three distinct layers. The lowest, and therefore oldest, layer was made of limestone packed with the fossilised remains of tiny one-celled creatures called *foraminifera*. The top, and thus most recent, layer was again limestone. Sandwiched between was a half-inch-thick layer of dark clay. But close inspection of these upper two layers revealed something rather odd: both were almost completely devoid of fossils. Walter's gift contained an apocalyptic message: 65 million years ago, something had happened to wipe out all but the tiniest *foraminifera*. That epoch – the so-called Cretaceous-Tertiary (K/T) boundary – is well known to geologists, but for another reason. It marks the mysterious end of the 100-million-year reign of the dinosaurs. Perhaps whatever wiped out the *foraminifera* killed the dinosaurs too.

What could have happened? The place to look for answers appeared to be in the composition of the curious half-inch dark clay layer at the middle of Walter's gift. His father, a swashbuckling scientist with a reputation for cutting through the small-minded conservatism typical of many academics, couldn't resist the challenge his son's gift presented. He decided he was going to solve the mystery of the death of the dinosaurs.

Reconstructing an apocalypse

The first task facing Alvarez was that of working out how long the clay layer had taken to form. This would give a clue to the nature of the catastrophe. If the layer had taken, say, 500,000 years to form, then the extinctions may have been the result of something like a gradual change in the Earth's climate. If it took 100 years, then something altogether more dramatic may have occurred.

Alvarez hit on the idea of measuring the rate of formation of the clay layer by exploiting the fact that a steady rain of meteoritic material falls on the Earth every day. Such material has a chemical 'signature' slightly different to that of the Earth's crust. In particular, it contains higher concentrations of the metallic element iridium. Assuming that the rate of meteorites coming into the Earth's atmosphere is constant, measuring the iridium concentration in the clay layer would give a measure of the rate at which the layer was formed.

Told of this idea by his father, Walter Alvarez passed on more samples of the Gubbio rock to Frank Asaro, a colleague and expert in using radioactivity to find the chemical composition of materials. In early 1978, the Gubbio sample was tested for iridium by exposing it to neutrons from a nuclear reactor. These cause any iridium present to give off a characteristic burst of gamma rays. After months of work, Asaro had some news for Luis and Walter: the clay layer did indeed contain iridium. The trouble was, it contained far too much to make sense. There was 30 times too much iridium in the thin layer of clay.

Richard Muller, a close colleague of Luis Alvarez who later played a part in developing his ideas (and wrote *Nemesis*, an enthralling account of the K/T boundary discovery), describes the reaction of those involved: 'There was no obvious explanation. Was it worth trying to figure it out? It could be just too hard a puzzle to solve . . . Luis felt differently. He thought the iridium excess was a new clue, unlike any that had been found before. He sensed a great discovery. Like a shark smelling blood, he sensed something worthy of attack, and he went after it with all his vigor. It was the clue that could unravel the mystery of the dinosaurs.'

A few months after Asaro had measured the iridium levels, Luis had indeed found an answer. The iridium, he believed, had come from the explosion of a giant star – a supernova explosion. This seemed to explain a lot: perhaps the death of the dinosaurs was due to the radiation and heat energy from the supernova which had blasted life on Earth. The supernova theory could be tested. Supernovae explosions are the furnaces in which all the heavy chemical elements in the universe were born. Thus, if the supernova theory was true, there were bound to be more telltale chemical 'signatures' locked in the clay layer.

Alvarez realised that among those signatures would be one for plutonium. Measurements of the level of a particular isotope of plutonium, Pu-244, would be powerful evidence that the supernova explosion had wiped out the dinosaurs. In March 1979, Asaro and his colleagues set about finding the plutonium. They soon succeeded. The Alvarezes were elated, and were planning how to break the news that they had discovered what had killed the dinosaurs.

But, in a curious repetition of Crick and Watson's mistaken early ideas about the structure of DNA, the Alvarezes were to be disappointed. Asaro had a reputation of being a very careful scientist and was not one to be rushed into headline-making announcements. Careful checking by his team turned up the depressing fact that, despite having taken very great care to prevent it occurring, the samples of clay had become contaminated by tiny amounts of man-made plutonium from an adjoining laboratory. The supernova theory was dead.

The rock of Damocles

Luis Alvarez was not down for long. If a supernova was not to blame, what else could dump so much iridium on the Earth, and wipe out the dinosaurs literally in one fell swoop? He turned to another potential source of iridium: the impact of an asteroid. Asteroids are huge chunks of rock, anything up to 1,000 kilometres across, which chiefly orbit the sun between Mars and Jupiter. They are thought to be the source of most of the meteorites that reach Earth – and are thus expected to be relatively rich in iridium.

Alvarez made some back-of-the-envelope calculations using data about the so-called Apollo asteroids, a class of asteroids whose orbits cut across that of the Earth. By estimating the mass of iridium locked up in the clay layer over the entire Earth, and assuming that asteroids and meteorites have similar proportions of iridium, he calculated that the asteroid that produced the layer would have to have a diameter of about five to ten kilometres. This was just the size range that astrophysicists predicted for an Apollo asteroid hitting the Earth around that time.

If the asteroid was roughly ball-shaped and made of rock, this size meant that the asteroid mass was roughly 10^{12} (a million million) tonnes. Sweeping into our atmosphere at more than 100,000 kilometres an hour, it would smash into the Earth with the energy equivalent of more than 10^7 H-bombs.

The results of an impact of this destructive energy beggar the imagination. Scientists are still calculating the likely results, but one thing seems certain: so huge a collision would throw a lot of debris – including iridium – into the air, and this would settle on the Earth hours, days, even months later. Alvarez realised he had probably found the explanation for the curious clay layer in his son's gift: it was, perhaps, the result of the impact 65 million years ago of a chunk of rock so vast that when the base struck the Earth, the top of it would be poking into space.

In June 1980, three years after Walter gave his father the gift, the prestigious American journal *Science* published the paper 'Extra-terrestrial Cause for the Cretaceous-Tertiary Extinction'. In addition to the evidence from Gubbio, the paper included the discovery of a huge iridium anomaly in a rock sample taken hundreds of miles further north, in Denmark.

The implications of the paper make it one of the most exciting and controversial ever written. But even Luis Alvarez's impeccable credentials could not protect him from the venom that followed his daring to claim a discovery in a field not directly related to his own. One fossil expert at Alvarez's own university branded the theory as 'codswallop'. Another described the work as a 'nutty theory of pseudoscientists posing as palaeontologists'. Like Alfred Wegener before him, Luis Alvarez discovered that scientists do not welcome outsiders wading in and discovering something interesting which they missed.

One early criticism was that the geological evidence showed that the dinosaurs died out over hundreds of thousands, perhaps a million years, and not in the space of a few weeks or years. But even the most careful examination of the best-preserved geological layers cannot pin down the time-scale of events to an accuracy of better than a few thousand years. Usual estimates are far less accurate. Thus it seems unlikely that any really convincing geological evidence for a sharp extinction event will ever be found. Some of Alvarez's supporters take comfort in this. Others, however, are not so sure. They have tackled the problem of a less sudden extinction by arguing that a whole swarm of asteroids – or comets – rained down on the Earth over many thousands of years.

Another important feature of the extinctions is that many of the species, including the dinosaurs, that disappeared around the K/T boundary appear to have been on the way out long before then. The Alvarez team now think that perhaps the impact provided the *coup de grâce* to creatures already on the brink of extinction.

Others, however, maintain that the impact theory is simply unnecessary. Perhaps something terrestrial such as an epoch of intense volcanic activity finished off the dinosaurs. Dust thrown high into the atmosphere could have blotted out the sun's light, killing off the plants on which dinosaurs and their prey relied. This might explain why the extinctions were spread out over so long a period. In fact in 1972, long before the Alvarezes had put forward their impact theory, it was pointed out that a period of great volcanic activity had indeed occurred about the time of the K/T extinction, with the most violent activity being centred on a region of northwestern India called the Deccan Traps. These huge lava flows, almost 2.5 km thick in places, attest to some devastating eruptions at the end of the reign of the dinosaurs.

Intriguingly, a number of geophysicists have recently shown that a huge plume of hot magma from the Earth's mantle was sitting right under the Traps 65 million years ago, and could have triggered violent volcanic activity. The same plume is now under Réunion, an active volcano in the Indian Ocean off Madagascar.

It may be that the two explanations of the death of the dinosaurs are linked, of course. Perhaps an impact into a thin part of the lithosphere triggered an era of intense volcanism. But the Alvarezes have defended their impact theory by saying that volcanoes cannot produce the chemical anomalies seen in the K/T layers. In particular, the ratio of various metals such as iridium and gold in air samples taken from volcanic eruptions often does not match those at the K/T boundary.

The evidence builds up

High concentrations of iridium have now been found at the K/T boundary at more than 80 different sites as far apart as Scandinavia and the Pacific. Minerals that have been subjected to huge jumps in pressure and temperature have also turned up in the clay layer of the K/T boundary.

But even more impressive is the evidence of the appalling consequences of the cosmic impact. In 1988, Dr Edward Anders of the University of Chicago and colleagues in the US, Switzerland and New Zealand published evidence for a fire which engulfed the planet 65 million years ago. They found huge amounts of soot in K/T boundary clay unearthed at five sites in Europe and New Zealand. Analysis revealed that the soot is the product of plants and trees that burst into flame. Much of the world's vegetation would have been laid low and made tinder dry by the shock wave and heat of the initial impact. The dust thrown up into the atmosphere by the impact would also lead to huge lightning storms. Perhaps these provided the spark for the flames which apparently consumed the whole planet.

It is now clear that if the initial impact did not kill off all the dinosaurs, the subsequent environmental catastrophes probably would. As well as iridium, asteroids are likely to be rich in nickel, a metal that is toxic to plant life. Thomas Wdowiak, an astrophysicist at the University of Alabama, believes that the entry of this metal into the food chain would have guaranteed the extinction of the dinosaurs.

But American scientists are now growing increasingly confident that they can point to the most impressive piece of evidence for the K/T impact: the huge crater it left behind. Early clues to the site of the impact came from studies of rocks in eastern Texas which suggested that some huge tsunami may have hit the US from an impact somewhere in the Gulf of Mexico. Other researchers found curious anomalies in magnetism and gravity in the area, suggesting that there was a huge buried crater in the area.

Then, in 1988, Alan Hildebrand of Arizona University discovered weird geological formations under the sea about 1,000 km south of

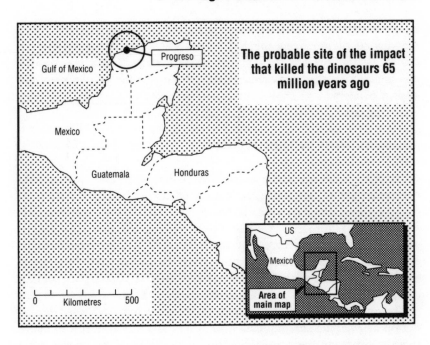

Cuba. Two years later, he revealed that analysis of rocks taken from the K/T boundary in Haiti showed that they contained high levels of iridium and other features expected from a huge cosmic impact somewhere in the Caribbean. Now Hildebrand is willing to point to what he believes is the most likely site of the impact: the northern coast of the Yucatan peninsula. A huge geological feature, circular in shape and centred seventeen kilometres east of Progreso, lies buried there. Known as the Chicxulub crater, it is 180 km across – just about the size expected – and many researchers now to accept that it is the site of the K/T impact. Walter Alvarez, in particular, says he is 'cautiously optimistic' that the site of the impact has at last been discovered.

Chicxulub may not be the only scar left behind by the terrible events of 65 million years ago. Geologists point out that the 36 kilometre-wide Manson crater buried in central Iowa, together with some features in Siberia, appear to date back to the K/T boundary. Piet Hut of the Institute for Advanced Study at Princeton believes the object that hit Chicxulub may have disintegrated before impact, creating other craters.

This would make sense if the impacts occurred during the summer in the northern hemisphere. Astonishingly, by studying the fossil remains of plants from the K/T boundary, Jack Wolfe of the US Geological Survey has managed to show that the Earth was indeed plunged into a dust-induced 'winter' in early June of the year the impact occurred.

Are extinctions regular events?

More than a decade after the original paper in *Science*, the debate over whether or not a single asteroid or comet impact could wipe out the dinosaurs rages on. It has now gone beyond just the K/T extinction 65 million years ago to encompass the last 500 million years of geological history. For the K/T event was far from being a one-off catastrophe. There is now evidence for at least five major extinctions of life, with one – about 225 million years ago – wiping out more than half of all marine life on the planet.

The powerful desire of scientists to come up with a unified explanation of events is leading to a fascinating array of theories to bring these extinctions under one theoretical umbrella. One of the most controversial aspects of these theories is whether or not there is any pattern in the timing of the extinctions. In 1984, American palaeontologists David Raup and John Sepkoski put forward evidence that the largest extinctions occur every 26 million years. Finding a regular interval in a mysterious physical phenomenon is often a powerful clue towards solving the mystery, and Raup and Sepkoski's work has led to several rival explanations for the alleged periodicity. Not that everyone is convinced there is anything to account for, however; many researchers believe that the evidence for regular extinctions is far from conclusive and may be just a statistical quirk.

Even so, some leading theoreticians have begun to look at how the Earth may suffer periodic apocalypses. One theory ties the extinctions to the movement of the solar system around our galaxy. The sun and planets complete a huge, circular orbit about the galaxy's centre once every 250 million years or so. But in addition, the solar system bobs up and down, passing through the thickest part of the galactic disc roughly every 30 million years. But the solar system is thought to be accompanied on its travels by a huge cloud of material – the so-called Oort cloud – from which comets are believed to come. Perhaps the Oort cloud is disrupted when it passes through the galaxy's disc, sending a hail of comets – and death – our way every 30 million years.

Arguably the most controversial explanation of the periodicity is that the sun has a stellar companion for its voyages around the galaxy. Aptly dubbed Nemesis by Richard Muller, the scientist and colleague of Luis Alvarez who proposed this theory in 1984, this as yet undetected star is presumed to be on a huge orbit around the sun, sweeping in once every 26 million years, disturbing the Oort cloud and triggering a comet shower on the way. By fiddling with the shape of the orbit, it is even possible to spread the comet shower out over a suitably long time to account for an epoch, rather than instant, of extinction.

Does the 'Death Star' exist? One problem is that we are currently (and thankfully) about halfway through the 26-million-year interval between extinctions. This means that, if it exists, Nemesis is a long way from us at present. It may also be a very faint dwarf star. Even so, Richard Muller is sure it will one day be found.

A sky-searching satellite called *Hipparcos*, launched in 1989, is sending back data which may one day locate Nemesis – if it exists.

Might we be wiped out by a comet?

Anyone learning of the dreadful events of 65 million years ago is bound to ask about the chances of the Earth being hit now. Certainly, impacts did not stop with the K/T event. Geologists have found evidence for well over a hundred huge impact craters formed over the last half-billion years. There is, in fact, a preponderance of more recent ones. This – fortunately – does not mean that the risk of obliteration is increasing. The trend towards younger craters is simply the result of erosion, which has obliterated all but the largest ancient craters.

But there is one place that has faithfully recorded every impact it has suffered over the last billion years or so: the moon. With no atmosphere or geological action to wipe the slate clean, the moon constitutes a 'ledger' of the debris that has been threatening life on Earth. Geologists have learnt to read this ledger, and they have found, happily, that the frequency of impacts is in fact the exact opposite of that suggested by studies of the Earth. For the first billion years of its existence, the Earth appears to have been subjected to an appalling battering. Indeed, a growing number of scientists now think that the 'splash' of incandescent rock created by the impact of one particularly huge chunk of debris, roughly the size of the planet Mars, may have created our moon.

In view of such primordial violence, it is not surprising that there is no evidence of life before about 3.5 billion years ago. But, ironically, it now seems that that violence may have played an important role in the birth of life. Some of the water now in the oceans emerged from minerals which gave up their water content as the Earth cooled. But in the last few years, researchers have been attracted by the idea of another source of water for the oceans: comets, whose icy contents were dumped and melted into water on impact. In the earliest days of the Earth, the water did not remain liquid for long; the continual bombardment by rocky debris simply evaporated it. But over the first billion years, the bombardment eased sufficiently to allow a net accumulation of water. Calculations published in 1990 by Dr Christopher Chyba of Cornell University suggest that as much as two thirds of all the water now on Earth may have been brought to us from comets during this period. Perhaps some of the basic ingredients for life – amino acids, or self-replicating proteins – were also brought to Earth by comets.

As the aeons passed, the rate of impact decreased steadily and life on Earth flourished. The work of Alvarez and others strongly suggests, however, that it has not been plain sailing ever since – every now and again (perhaps even on a regular basis), something has loomed out of the vast darkness of space to put the march of evolution on a completely new track. Without a K/T extinction, dinosaurs might well still be reigning supreme.

Could we be struck a devastating blow from the heavens today? Some researchers claim that this is more than just a possibility; it may have happened at least twice in recorded history. Dr Victor Clube, an astrophysicist at Oxford University and one of the leading figures in what might be called the 'cosmic catastrophism' movement, believes that part of a comet struck the southern counties of England around the year 441. Clube's evidence comes from an unusual source: an Anglo-Saxon chronicle written by a cleric called Gildas. He talks of Britain suffering a major catastrophe around this time, and makes explicit reference to a 'fire that fell from heaven' which 'did not die down until it had burned the whole surface of the island'. Dark skies and huge migrations out of the area followed, with the land still being in ruins a century after the disaster.

The conventional interpretation of Gildas's statements is that they are metaphorical accounts of the invaders that ravaged Britain following the fall of the Roman Empire. But Clube thinks that Gildas should be taken at his word – perhaps the 'Dark Ages' are more than just a gap in the historical record, and really were dark.

Impressive support for Clube's view lies in ancient records kept by Chinese astronomers. These make reference to the appearance of a strange comet in the same year as Gildas's 'heaven-sent' catastrophe. This has led Clube to name as the likely culprit a comet still visible in the last century: Biela's Comet.

Discovered by the eponymous Austrian amateur astronomer in 1826, Biela's Comet has a history of bizarre behaviour. In 1846, it was seen to split into two, with the two fragments appearing together in 1852. But then it disappeared. In its place twenty years later a shower of meteors was seen, apparently signalling the final destruction of the comet into small chunks of rock and ice which burnt up spectacularly – and harmlessly – in the atmosphere. Working backwards from the nineteenth-century observations, Clube has discovered that Biela was very close to the Earth in the fifth century. Furthermore, he has found evidence that showers of meteors – shooting stars – were being produced by the comet even then, suggesting that Biela was already unstable. Perhaps, argues Clube, Gildas was giving an account of the entry into our atmosphere of a huge chunk of the comet, which had survived its journey through the atmosphere and smashed into Dark Age Britain. The absence of any obvious crater in the southern counties suggests that the relatively weak cometary material actually disintegrated before striking the ground. The resulting blast would still have been devastating.

Five hours from catastrophe

Historians will no doubt continue to argue about Clube's interpretations of the chronicles. But it now seems clear that in our own century we have been within a matter of hours from a major cosmic catastrophe. On the morning of 30 June 1908, peasants living in a remote part of northeastern Siberia witnessed an event that literally shook the world. A huge fireball, too bright to look at directly, appeared in the cloudless sky. It was accompanied by a deafening roar, and by vast plumes of black smoke spewing out in its wake. As the fireball approached the ground it appeared to break up, and a colossal explosion was heard. Then there was silence, broken only by the weeping of women who thought that the end of the world had come.

A meteorologist working near the Siberian village of Nizhne-Karelinsk set about trying to locate the point of impact of the object. Eyewitness reports indicated a region close to the Tunguska river. The sheer force of the explosion then became clear. A farmer working 40 miles away from the impact point was thrown several feet by a shockwave produced by the fireball. Houses 200 miles away were shaken; the explosion was heard 500 miles away. Seismometers detected the reverberations of the explosion circling the entire planet. A contemporary issue of *The Times* carried a letter from golfers in the southeast of England who said that following the explosion the night sky was so bright they could have completed a round at two o'clock in the morning.

But what could have caused so huge an explosion? The difficulties of getting out to the site of the 'Tunguska event' delayed the start of a search for its cause for almost a generation. Finally, in March 1927, a Soviet mineralogist named Leonid Kulik set out from Tayshet junction on the Trans-Siberian Railway and headed north towards the Tunguska river. Kulik had the backing of the Soviet Academy of Sciences and he was determined to prove his own theory that a huge meteorite had smashed into the Earth.

Within six weeks, Kulik and his men reached the river Mekirta, about fifteen miles or so south of the impact point. Now, looking north, they saw a scene of utter devastation. Thousands upon thousands of trees had been flattened, their trunks scorched and their tops all pointing to the southeast. Kulik's guides refused to take him any further, forcing him to return to the nearest major settlement for new companions. But by June, Kulik was back, following the lines of trees to the very centre of the devastation. At latitude 60° 55' north, 101° 57' east, he found himself in a shallow, mile-wide depression. Looking out from its centre, all the trees pointed outward. He had found the very heart of the Tunguska event.

Kulik immediately set about trying to find fragments of the meteorite he thought had caused the devastation. However, both on this

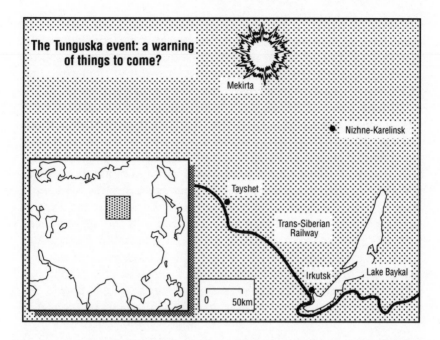

The Tunguska event: a warning of things to come?

Mekirta

Nizhne-Karelinsk

Tayshet

Trans-Siberian Railway

Irkutsk

Lake Baykal

0   50km

expedition and on his second, third and final returns to the site, he failed to find anything. In 1942, Kulik died defending Moscow against the Nazis. By then, his surveys had filled in some details of the event, including the trajectory of the object through the Earth's atmosphere. But of the meteorite, he found no trace.

It is now thought that Kulik's lack of success stemmed from the fact that he was only partly correct. Something had indeed struck the Earth, but a stone or iron meteorite would have left at least some fragments behind. However, comets are thought to be far more delicate objects, with a density similar to that of pumice; a chunk of comet would therefore probably not have withstood the aerodynamic buffeting it got as it hurtled into the denser parts of our atmosphere. Thus the final, huge explosion heard by the peasants may well have signalled the destruction of the comet fragment five miles up, rather than its impact on the ground. It seems we were decidedly lucky with the timing of the break-up. If the Tunguska object had struck five hours later, it could have wiped out Helsinki with the effectiveness of a fifteen-megaton H-bomb.

As with the Dark Age comet, astronomers have investigated the possibility that the debris came from a comet which still orbits the sun today. In 1947, British astronomers using radio telescopes announced the discovery of a daylight shower of meteors emanating from the direction of the constellation Taurus. These meteors follow an orbit round the sun which cuts across the orbit of the Earth. The astronomers found that the peak of the Taurid meteor shower occurs on 30 June – the very same day of the year as the Tunguska event.

Scouring the catalogues of comet orbits, it was then discovered that the orbit of these daylight meteors is very similar to that of a comet named after the nineteeth-century German astronomer Encke. In its travels, Encke's Comet passes close to the sun, almost as close as the planet Mercury, before swinging out again into the cold, dark space beyond the asteroid belt. As a result, Encke's Comet experiences temperatures hot enough to melt tin and cold enough to liquefy nitrogen. This is likely to make it particularly unstable. Perhaps the Tunguska event was caused by a particularly large piece of Encke's Comet that broke free.

But recent research by Clube at Oxford suggests that Encke, the Tunguska object and the Taurid meteors may all in fact be fragments of a single giant comet, perhaps 50 miles or more across, which smashed into the asteroid belt between Mars and Jupiter about 5,000 years ago.

Can we do anything to protect ourselves from cosmic devastation? Following Alvarez's work on the K/T impact, the US National Aeronautic and Space Administration commissioned a report on what might be done in the event of an asteroid or comet being spotted soon enough to take avoiding action. This in itself is unlikely – telescopes with the power to detect the feeble reflected light of a tiny piece of comet have very small fields of view. That is, they cannot take in very large regions of the sky in one session. Setting up an effective detection system would be very expensive.

Supposing astronomers were able to detect a chunk of cosmic debris early enough, the obvious thing to do would be to attack it with nuclear warheads. However, this would not work with a comet or asteroid of typical size, ten kilometres or more across. Even if the largest fragments were just one hundredth the size of the original, we could still expect Tunguska-like devastation.

The NASA committee concluded that the best strategy would be to send a probe to the object, dig a hole in its surface and set off a nuclear explosion within it. The blast would then – hopefully – divert the thing away from Earth.

Soviet scientists, perhaps not surprisingly given the size of their country and the fact they have already been hit, have taken a particular interest in avoiding a repetition. In 1989, a team from the Central Aerohydrodynamic Institute in Moscow put forward what is certainly an imaginative way of protecting the Earth. Instead of moving the comet, they propose moving the Earth.

As the Earth and moon are gravitationally bound to one another, a change in the orbit of the moon can affect the Earth. Calculations presented to the International Astronautical Federation by the Soviet team suggest that destroying about 0.2 per cent of the moon would enable the Earth to dodge an incoming comet by 10,000 kilometres or more.

Are such drastic measures justified? The lunar cratering record suggests that Tunguska-sized objects may strike the Earth as often as once every 500 years. We may therefore have some time to work out a realistic strategy – but not an indefinite amount of time. Indeed, we may already be able to see the object that will wipe out the human race unless we find a way of defending ourselves. In 1977, American astronomer Charles Kowal discovered a huge object over 300 km across orbiting between Saturn and Uranus. Called Chiron, it is now known to be a comet. Its orbit is unstable and its precise behaviour cannot be predicted with certainty. Perhaps Chiron is simply biding its time before setting off on one last plunge into the heart of the solar system – with the Earth as its final destination.

OUTSTANDING MYSTERIES
How well can we predict natural events?

The hallmark of a scientific theory is that it makes a testable prediction. The better a prediction fits the facts, the more credible the theory becomes. For centuries it has been an article of faith among scientists that if you have a very good theory and a decent amount of data, you should be able to make such predictions as accurate as you like. But over the last thirty years or so, researchers have been studying phenomena which wreck so simple a view. They have discovered that it is possible to know all the laws governing some phenomenon and yet still not be able to predict with any accuracy what will happen.

Why has it taken scientists so long to learn that there are basic limits to what they can know? The reason lies in the way they traditionally attack complex problems. The usual trick is to make a problem less complex by throwing out everything except what is judged to be the essence of the problem.

This might seem a perfectly sensible thing to do, and indeed it worked well enough for centuries. But in simplifying the description of nature, scientists got into a habit of throwing away something that we now know is crucial to the description of many natural phenomena: the so-called 'non-linear' features. When these act in concert, their overall effect is more than merely the sum of their individual effect. They influence each other to produce total effects of baffling complexity.

This sounds complex, but its essence is familiar to us all. Suppose we are going on a trip abroad. On the morning of the trip, we oversleep by fifteen minutes. We rush around, and eventually leave the house just five minutes late. Unfortunately, we still miss our train to the airport, and thus miss our flight; the next is not until the following day. Thus by sleeping for just fifteen minutes more than we should we end up not fifteen minutes late at our holiday destination, but a whole day late. This is the essence of non-linear situations: a small change can produce drastic effects.

Trying to investigate the end result of non-linear influences on a problem using 'pen-and-paper' mathematics is hideously difficult; hence the keenness of scientists to throw out non-linearities wherever they reared their heads. But by the 1960s, a way to handle the full complexity of non-linear phenomena had emerged: the computer. These electronic drudges were willing to do without complaint all the donkeywork of following the complexities of non-linear phenomena.

Scientists using early computers soon realised that by ignoring non-linearities they had been restricting themselves to a surprisingly dull picture of what the world is really like.

Everyone fancies themselves as weather forecasters and in the early 1960s Edward Lorenz at Massachusetts Institute of Technology set about building up a crude mathematical model of what he thought a decent theory for the weather should involve. By feeding in an initial set of data, simulating the sort of temperature, wind speed and other information that would come in from weather ships and elsewhere, Lorenz had the idea of setting the computer to predict the weather to follow.

Lorenz was particularly interested to see what would happen if he asked the computer to make longer- and longer-term forecasts. His computer – a Royal McBee LGP-300 – was incredibly sluggish by today's standards, capable of only 60 multiplications a second. So, to avoid having to start from scratch every time, Lorenz decided to push his forecasts out further by using the values it had reached at the middle of its original run as the starting values of the next.

When he set the machine off again, he was expecting the computer faithfully to repeat the second half of the first run before producing the new, longer-term prediction. Lorenz went off for a coffee while the machine started up again. But when he returned, he discovered that the computer had a surprise for him. Although it had indeed reproduced the first run to start with, the computer was soon producing entirely different results. Lorenz's first reaction was to blame a faulty valve in the machine. But then he realised what the real cause of the trouble was. The computer's memory stored numbers to six decimal places. However, to save space, it printed them out only to three. When starting the computer off on the second run, Lorenz had fed in these truncated values, never thinking that the difference – just one part in a thousand – would have so great an impact.

Lorenz was wrong – and was on the brink of a major discovery. The equations he was using to make his forecasts contained non-linear terms. As we have seen, if such a non-linear system is changed even slightly, the consequences can be dramatic. Lorenz gave this key feature of non-linear systems a name: the Butterfly Effect. It reflected the discovery that, in principle at least, the beating of a butterfly's

The Butterfly effect

How two forecasts with similar starting conditions can be forced off track after 10 days by chaotic phenomena – making long–term forecasts impossible

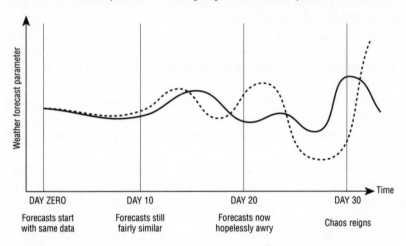

DAY ZERO	DAY 10	DAY 20	DAY 30
Forecasts start with same data	Forecasts still fairly similar	Forecasts now hopelessly awry	Chaos reigns

wings in America could affect the weather system enough to trigger a typhoon in Hong Kong.

Lorenz dug further with his little computer, and investigated the relatively simple – albeit non-linear – phenomenon of convection, in which hot air rises, cools and falls again in huge loops reaching up into the atmosphere. Even with a computer, the equations governing convection are complex and so Lorenz did the usual trick of throwing away most of the mess. But he kept the parts that represented the way in which the various key features influenced one another. These stripped-down results he fed into his computer to see how the equations behaved as the various quantities in them were altered.

Lorenz's computer solved the equations by making a guess at the answer and then feeding the result back in to get a more accurate answer. His computer was capable of making only one guess every second, so it took an hour to get a decent set of results.

They proved well worth the wait. Plotting the outcome of each guess on graph paper, Lorenz discovered that, to start with, the numerical value of the guesses wobbled in a periodic manner, with the size of the wobble increasing with each guess. But then, after about the 1,600th guess, the computer started to churn out completely wild results. The wobbles were no longer periodic; they showed no pattern at all. They were *chaotic*.

By careful analysis of his results, Lorenz did detect some underlying pattern amid the chaos. But his attempts to extend the short-term predictions far into the future turned out to be doomed by the Butterfly Effect.

Are long-term weather forecasts feasible?

At the end of his 1963 paper publishing his findings, Lorenz talked about their implications for long-range weather forecasting. They make salutary reading. 'Prediction of the sufficiently distant future is impossible by any method, unless the present conditions are known exactly,' Lorenz concluded. 'In view of the inevitable inaccuracy and incompleteness of weather observations, precise very-long-range forecasting would seem to be non-existent.'

But how long is 'very long range'? How long does it take for the butterfly's wings to make their influence felt? Lorenz could not answer precisely, but said that it could be anything from days to centuries. Now, using some of the world's fastest computers, capable of thousands of millions of calculations a second, meteorologists are trying to find out just how far they can push their predictions.

Meteorologists are confident that predictions as much as five days ahead can be made with considerable accuracy, especially in temperature. The question surrounds the accuracy of forecasts from five to ten days ahead, and beyond.

To investigate this question, Dr Tim Palmer, head of the predictability section of the European Centre for Medium-Range Weather Forecasts in Berkshire is investigating the problems of long-term forecasting using a whole family – an 'ensemble' – of different starting points. Palmer's idea is to begin each prediction with a slightly different set of initial conditions, such as temperature, wind speed and direction over America. He can then see what happens as the prediction goes further into the future.

For some starting conditions, each of the long-term predictions produced by an ensemble are pretty much the same, even after a week or so. But for others, predictions start to become very different after no more than five days. In such cases the ensemble of initial starting points constitutes a far less stable weather pattern, one which is far more susceptible to the influence of the proverbial flap of a butterfly's wing.

It turns out that some parts of the world, and some times of the year, make for more unreliable predictions. It will come as no surprise to Britons, for example, to learn that the weather in our latitudes is particularly unpredictable, especially in winter. Palmer thinks that even if precise prediction of the weather ten days ahead proves elusive, it may be possible to use the ensemble approach to put some reliability figure on a weather forecast. This figure would get lower the further out the prediction is pushed, and the more unstable the weather systems involved in starting off the prediction.

Such 'reliability analysis' is some years off, however, because even making one ten-day prediction of the weather over Europe pushes current computer technology to the limit. Running whole ensembles for every forecast would require much more power, and must await the next generation of 'hypercomputers', capable of a million million calculations a second. Such machines are now under development in America, with funding from the Pentagon.

Seeing into the far future

But if weather forecasting is always going to be limited to no more than, say, two weeks ahead, how can anyone possibly predict the climate of the whole Earth for the next century? Such calculations are crucial to our understanding of the impact of global warming, whose effects take decades to show up. But does the Butterfly Effect wreck any hope of sensible predictions?

It is here that the distinction between weather and climate becomes crucial. Weather is, roughly speaking, a short-term manifestation of climate. The weather determines if it will rain tomorrow morning. The climate determines if it will rain more in certain months of the year than in others. So, although weather can appear pretty random over a timescale of hours to days, over a timescale of years to decades it will follow a more or less settled pattern. This pattern of weather is the climate. Meteorologists keen on chaos theory talk of the climate as a meteorological 'attractor': a stable state to which the weather is 'attracted'.

Climatic prediction is thus concerned with the behaviour of this attractor, rather than the little patterns that make it up. One of the big questions in current climatic research is whether or not global warming is changing the nature of the attractor – that is, changing the whole climatic system of the Earth.

Many countries are experiencing freak weather conditions which armchair pundits put down to global warming. It is entirely possible, however, that the extremes are part of the normal climatic attractor which has reigned for centuries.

Then again, they may not be. The mathematics of chaos allows for the climatic attractor suddenly to flip into a completely different form if disturbed badly enough by outside influences. Industrial pollution, which locks in more of the sun's heat, could potentially destabilise the climatic attractor. If the attractor decides to 'flip' suddenly, we would be in dire trouble. Growing seasons could be turned on their heads as mild winters are followed by dismal summers. Deserts could bloom again, while lush pastures are turned into arid dust bowls.

Could we wake up tomorrow to a world in the grip of a new Ice Age? Fortunately, the world's oceans should prevent that. The huge amount of heat energy needed to change their temperature significantly – their 'thermal inertia' – is sufficient to rule out really rapid changes in climate. Climatologists usually talk of a century or so being needed for substantial changes in the climatic attractor to become noticeable.

However, recent work by Dr Willi Dansgaard at the Geophysical Institute of the University of Copenhagen suggests that we may not be so well protected from the effects of a sudden attractor flip. By studying ice cores taken from Greenland, Dansgaard discovered that the last major change in the Earth's climate – the ending of the last Ice Age 10,700 years ago – did not take place over centuries, but in a matter of decades. In the space of just 50 years, the average annual temperature of South Greenland soared by 7°C.

The rise of just a few degrees predicted to occur in the coming decades as the result of industrial pollution has been enough to trigger worldwide concern. The effect of a 7°C rise can only be guessed at.

Techniques similar to those being developed by Tim Palmer for weather prediction are now being used for climatic prediction too, to see if such dire scenarios are likely. The results are still not in. Until they are, any assertions about the amount of time left to take action to protect our environment should be viewed with much scepticism. We may have much less left than we think.

Chaos in the majestic clockwork

The idea of chaos lurking at the heart of weather forecasting will probably not come as so big a surprise to cynics who think most forecasts are no better than flipping a coin in any case. But there are natural phenomena, long considered to be paragons of predictability, that are now revealing themselves to be equally fickle.

The stately movement of the planets around the sun has for centuries seemed like some vast, majestic clockwork, propelled by gravity. The eighteenth-century mathematician Pierre Simon de Laplace felt moved to make the following pronouncement about the nature of the solar system, indeed the entire universe: 'An intellect which at any given moment knew all the forces that animate Nature and the mutual positions of the beings that comprise it, if this intellect were vast enough to submit its data to analysis, could condense into a single formula the movement of the greatest bodies of the universe and that of the lightest atom: for such an intellect nothing could be uncertain; and the future just like the past would be present before its eyes.'

This has a certain ring of arrogance about it, and indeed there is now considerable evidence that such statements are mere exercises in hubris. The evidence originates in attempts to do precisely what Laplace was describing – predict the past and future state of the solar system.

Newton's law of gravity tells us that if we have just two objects in the whole universe – say, the Earth and the sun – we can predict both the past and future movements of the one about the other. But besides the Earth there are eight other planets going round the sun, each with mass and thus gravitational influence. What happens when we take their influence on the orbit of the Earth into account? Do all the planets still behave themselves, trundling around the sun for billions of years? Or do the mutual interactions of the planets make the majestic clockwork go haywire? Laplace himself looked at the question of the long-term stability of the solar system. And, of course, possessing only quill and paper, he was forced to make some simplifying assumptions. Sure enough, he fell straight into the trap of throwing away the very terms which, it is now known, play a key role in determining the stability of the solar system. As with the stability of weather predictions, it required the advent of powerful computers to make a serious study of the question feasible.

The first major computer-based attempt to find out if the solar system is stable was made in the early 1980s by an international team led by Professor Archie Roy of the University of Glasgow. Called Project LONGSTOP (for LONg-term Gravitational STudy of the Outer Planets), the idea was to look at how the mutual gravity of the planets, from Jupiter outward, affected each other's orbits over 100 million years. A Cray supercomputer, capable of hundreds of millions of calculations a second, was put on to the job, and the team waited to see what turned up. The result was rather boring – there seemed no signs of anything particularly wild happening as the millennia rolled by.

But then some researchers wondered whether the LONGSTOP studies had been long enough; maybe if the calculations were pushed out even further, something interesting would show up. The problem was that something even faster than a supercomputer would be needed to find out.

Fortunately, a team at the Massachusetts Institute of Technology had been working on this very difficulty. Their answer was the Digital Orrery, a custom-made solar system computer. It is named after the mechanical model of the solar system commissioned by the Earl of Orrery in the eighteenth century. Unlike an all-purpose supercomputer, the Digital Orrery has been specially wired up so that all its power is spent carrying out the calculations needed to predict the motion of the planets hundreds of millions of years into the past or future.

In 1988, the Digital Orrery was set to work on finding out the future fate of the outer solar system. The calculations were pushed out far beyond those of the LONGSTOP project, to more than 800 million years into the future. The results revealed that LONGSTOP did indeed miss something. For a start, something odd started to happen to the orbit of the planet Pluto, on the outer fringe of the known solar system. Pushed and pulled by the gravity of the other planets, Pluto showed signs of having a chaotic orbit whose behaviour could not be predicted with certainty for all time.

Newton's law of gravity states that every object in the universe affects every other, so any chaotic behaviour even by Pluto can, in theory at least, produce effects on all the other planets. So is the Earth following a chaotic orbit? The question is not an academic one. In 1920, the Serbian scientist Milutin Milankovitch showed that changes in the orbit of the Earth have the ability to trigger ice ages, by altering the concentration of the sun's heat reaching the planet.

Finding chaos in the Earth's orbit is far from simple. For every orbit Jupiter makes around the sun, the Earth makes almost twelve. Tracing the movements of the inner planets demands a lot more calculations and thus computer power. Despite the difficulties, Dr Jacques Laskar of the French Bureau des Longitudes in Paris carried out some preliminary calculations of the stability of the inner planets, including the Earth. Like Laplace, he was forced to make the problem manageable by throwing out some of the complexity, but was able to retain enough to draw some useful – and disturbing – conclusions. In 1989, he published results in the journal *Nature* demonstrating that the orbits of the inner planets do indeed show signs of being chaotic. The amount of chaos is quite high: the positions of the inner planets become effectively unpredictable in just a few tens of millions of years. Laskar found that the mutual gravitational influence leads to a powerful Butterfly Effect: an error in the distance of the Earth to the sun of a mere fifteen metres is enough to ruin any hope of saying where the Earth will be 100 million years later.

Laskar's discovery of chaos within the majestic clockwork of the solar system does not mean that the Earth could go wandering off into space at any time. However, it could have major implications for the future climate of the Earth – and thus for the future of the human race.

How was the solar system formed?

The discovery of chaos in the motion of the planets shows that there are fundamental limits to our ability to predict the future state of the solar system. But what of its far past – in particular, its origin?

How the solar system was formed is a mystery that has been attacked, off and on, for more than 200 years. Like many fundamental questions, the majority of scientists don't fret about it. But, as ever, answering it is more than a matter of tying up academic loose ends. It has a key bearing on the even greater mystery of whether or not we are alone in the universe. For if the conditions needed to bring about a complete solar system are relatively easy to achieve, there may be a myriad other planets in our galaxy – and some of those may be inhabited.

Cosmogony, as the study of the origin of the solar system is called, had some curious origins of its own. One of its pioneers was the eighteenth-century German philosopher Immanuel Kant, someone more usually linked with philosophy than physics. Though never venturing far from his home town of Konisberg, Kant's thoughts took him deep into the cosmos to ponder the origin of stars and galaxies. He was well qualified to conduct such a voyage of the mind, having studied mathematics and physics at Konisberg University.

In 1755 he published the results of his deliberations in a treatise called the *General History of Nature and Theory of the Heavens*. In it, he suggested that the band of light seen stretching across the night sky, the Milky Way, was part of a huge lens-shaped collection of stars. This turned out to be a remarkably prescient idea: only in this century was it finally proved that the Milky Way is on the huge spiral arm of the galaxy in which we live. Kant also predicted that the Earth was being slowed down by the drag of the Earth's oceans over its surface during tides – again, right on target.

And he made one other major prediction: that the solar system condensed out of a huge nebula of gas and dust. This idea languished for more than forty years until the brilliant, if somewhat unpleasant, French mathematician Pierre Simon de Laplace put forward a similar idea, probably independently, in his *Essay on the System of the World* in 1796. Bringing his far greater understanding of Newton's laws of mechanics to bear on the problem, Laplace produced a far more detailed theory.

It starts with the idea of a huge hot, dusty and spherical nebula slowly rotating on its axis. As it cooled, the nebula started to contract under its own gravity. At this point a law crucial to all theories of the origin on the solar system came into play. It is called the principle of conservation of angular momentum. The most familiar demonstration of the principle in action is the way in which spinning ice-skaters spin faster when they bring their arms over their heads. If they want to slow the spin, they simply open their arms out again. There is a relationship between the rate of spin and the extent to which their arms are extended – and it follows from the conservation of angular momentum: the smaller the extension, the faster the spin.

In just the same way, Laplace envisaged the shrinking nebula starting to spin faster as it got smaller and thus starting to alter its shape. Instead of being a sphere, the nebula starts to bulge around its middle, becoming more ellipsoidal as the collapse progresses. Laplace claimed that as it shrank the nebula spasmodically shed rapidly rotating material from its outer edge, leaving rings of dust in its wake. The material inside these rings itself shrank, said Laplace, until clumps of material started to form. It was these clumps which went on to form the planets we now see, with the sun at the centre of it all being the last member of the solar system to condense out of the nebula.

Now if Laplace's theory was even vaguely correct, the planets should all orbit in the same direction as the sun's rotation, and follow orbits lying roughly on the same plane as one another. The first feature is certainly true and the second, given a bit of licence and maybe a bit of chaos over the millennia, is more or less true. Laplace's nebular hypothesis, as it became known, thus seems a very neat theory and indeed it enjoyed great popularity with the post-Newtonian scientists of the nineteenth century. However, it suffers from one simple yet insuperable flaw. The sun contains 99.9 per cent of all the mass in the solar system. That means that it should also carry the bulk of the angular momentum of the solar system. Yet the mass and rotation rate of the sun have been measured, and account for just 3 per cent of the angular momentum. So where did all the angular momentum go? Put another way, how did the collapsing primordial cloud produce such a huge, yet slowly rotating object – the sun – at its centre?

Many scientists have thought it a shame to ditch so neat an idea because of this technicality and have made efforts to patch up Laplace's theory. The patch-ups continue to this day. One obvious way out of the angular momentum problem is to separate the formation of the sun from that of the planets. The primordial cloud might, for example, have been picked up by the sun on its travels around the galaxy. It could then collapse around an already existing sun, which could spin however it fancied.

Encounters of the primordial kind

But the sad fact is that angular momentum is not the only problem facing those trying to account for the origin of the solar system. Trying to get a planet to condense out of the extremely thin, cold material thought to make up the raw materials available is far from easy. Calculations show that to get a planet the size of Neptune to condense out of a nebula would require a period something like twice as long as the solar system is known to have existed!

Some astrophysicists, notably Professor Michael Woolfson of York University, think that it is time to stop trying to patch up the Nebular Hypothesis. They are advocates of a theory which can trace its origins back ten years before Kant's musings.

In 1745 the French naturalist Georges de Buffon suggested that the material for the planets had been smashed out of the sun by the passage of a comet. Laplace knew of Buffon's theory and his criticisms and alternative theory effectively buried it for more than 150 years. But in 1916, the distinguished English astrophysicist Sir James Jeans revived a variation on Buffon's ideas with what was to become the Nebular Hypothesis's most popular rival. Instead of a feeble comet – which we now know would have nothing like the influence needed – Jeans postulated the passage of a star close to the sun. The Encounter Hypothesis was born.

Jeans proposed that as the star approached, it dragged huge amounts of material from the sun's surface in a long filament. The filament then cooled and broke up, condensing into a string of lumps which went on to form the planets.

By the 1930s, this theory was in trouble as well. Although it did not have to worry about explaining the low angular momentum of the sun, the theory had problems explaining why the planets orbited where they did. And there was a second, far more serious problem. Measurements of the levels of light chemical elements such as lithium and beryllium in the Earth's crust strongly suggested that the Earth was never once part of the sun.

Jeans himself admitted that there was a lot wrong with his ideas. But its death throes produced a further account of the origin of the solar system – and one that has yet to achieve the recognition it may well deserve. This is the so-called capture theory of Professor Woolfson and his long-time collaborator Dr John Dormand of Teesside Polytechnic. Unlike Kant, Laplace or even Jeans in this century, they have been able to simulate in great detail events such as the passage of stars by the sun, using computers.

The picture that emerges is as follows. The sun was born in a stellar cluster surrounded by other stars. This is certainly not unreasonable and ties in with what is now known about the birth-places of stars. After being formed, the sun *captured* a filament of material from one of its neighbours, still in the process of being born. The material then condensed into the planets, whose initially wild orbits became circular through the resistance offered by the debris left over from the planet formation process.

Two major problems are immediately overcome by the capture theory. First and foremost, as the theory does not try to link the origin of the sun to that of the planets, it doesn't have the angular momentum problem that scuppered Laplace's ideas. Next, coming from an as yet unformed star, the filament material would be rela-tively cold, getting rid of the problems, such as accounting for the presence of light elements on the Earth, thrown up by a hot-gas origin.

Woolfson and Dormand have, however, pushed their theory further, to consider the details of what would be formed out of the filament. It turns out that only bodies like the 'gas giants' Jupiter, Saturn and Uranus can be formed directly out of the filament. To account for the existence of the rocky planets such as the Earth, Woolfson and Dormand have postulated the existence of two more gas giants, rather unimaginatively called A and B. To account for the inner, Earthlike planets, Woolfson and Dormand claim that A and B destroyed themselves in a collision during the earliest days of the solar system, when all the planets were following their initially wild orbits. Gas giants are thought to have solid cores, and it is from the cores of A and B that the inner planets are said to have formed.

The Earth and moon, then, were once deep inside some huge planet wrecked over four billion years ago. Analysis of some of the 380 kilograms of moon rock brought back by the Apollo missions tends to support this view. Woolfson and Dormand have also used the capture theory to account for the formation of satellite systems, comets and much else besides. Added together, its explanatory power is impressive – especially compared to its rivals.

Yet the capture theory has yet to capture a major following among astronomers. This is another demonstration of what one could call the sociology of science. Much of the current work on solar system physics is done in the United States, which funds the planetary probes such as Pioneers 10 and 11 and Voyager 1 and 2. The capture theory, with all its messy collisions and primordial upheaval, is not in fashion there. At the end of their book *The Origin of the Solar System* – effectively a manifesto for capture theorists – Woolfson and Dormand make clear they know what they are up against, and make a plea: 'Any theory, no matter how convincing it may seem at any particular time, may be negated by new evidence with which it is totally incapable of being reconciled. Any theory worth taking seriously must be vulnerable. The capture theory is certainly vulnerable and, for the time being at least, is worth taking seriously.'

Now that the idea that the solar system is subject to violent events – such as the impact which may have killed the dinosaurs – is no longer seen as the preserve of lunatics, the day when the capture theory becomes the favoured bandwagon (or paradigm, as scientists prefer to call it) for the origin of the solar system may not be too far off.

Are we alone in the universe?

The American astronomer Carl Sagan once observed that however the answer to this question may turn out, the implications are profoundly disturbing. As he put it in a research paper written with Soviet astrophysicist I.S. Shklovsky in 1966, 'Finding life beyond the Earth – particularly intelligent life – wrenches at our secret hope that Man is the pinnacle of creation.'

This might explain why the debate on extraterrestrial intelligence (ETI) has been blighted by feeble argument and sheer irrationality ever since its recorded origins in ancient Greece. Democritus, Heraclitus and Epicurus – among others – argued for the existence of other inhabited planets. Their reasoning was based on the view that the universe must be very large and anything that can exist in such a universe does exist.

Such an argument is unlikely to impress many today – although it is undoubtedly the case that in an infinite universe there must be life – indeed a literally infinite variety of it. But weak argument is not the only aspect of the debate to emerge with the Greeks. The supercilious tone so often adopted by those against the idea of ETI is another. Plato considered anyone who did not hold our solar system to be unique to have 'a sadly indefinite and ignorant mind', and felt certain the Earth was unique in having life upon it. Aristotle's placing of the Earth at the centre of all things philosophically ruled out any question of life elsewhere. In later centuries his views acquired the status of dogma, with questioning punishable by death – hardly a conducive intellectual atmosphere for raising the level of the debate on the existence of life elsewhere.

But by the sixteenth century Aristotelian dogma had been undermined by the work of astronomers who had actually looked at the universe and seen how it really is. Most revolutionary was the work of the Polish astronomer Nicolaus Copernicus. The publication of his *On the Revolutions of the Celestial Spheres* in 1543 gave a much neater, more appealing account of the solar system and events within it by putting the sun rather than the Earth at its centre. But there was something much more important underlying the displacement of the Earth from the centre of all things – it knocked on the head the idea that the Earth is in some way special. Thus, Copernicus can be credited with reopening the ETI debate after centuries of repression. The problem since has been the sheer complexity of the factors involved in answering the question 'Are we alone?'.

The Drake Equation

In 1961, after centuries of wallowing around in a sea of personal prejudice and half-baked logic, an American physicist at Cornell University, named Frank Drake, came up with a small equation that gave the ETI debate something approaching a foundation on which to build. He arrived at the equation by bolting together some sensible guesses on what factors are likely to be important in predicting how much ETI exists.

Drake's Equation is basically this: the number of advanced civilisations in our galaxy is given by N, where $N = A \times B \times C \times D \times E \times F \times G$. The seven symbols, A to G, on the right-hand side of this equation are the heart of Drake's contribution to the ETI question. They represent the factors that are likely to be involved in the emergence and survival of an advanced civilisation in our galaxy. They may well not prove to be exhaustive – indeed, Drake's original equation was not as detailed as the version we give here – but they do provide a way of cutting through philosophical prejudice to get to the issues at the centre of the ETI question. By trying to estimate each of the factors in turn, we can also make a stab at answering the question of whether or not we are alone in our galaxy – if not in the whole universe.

We can make a start on fairly solid ground. The number of extant civilisations must depend on the number of new stars being born every year in the galaxy – if you have no stars, you have no planets either. Thus symbol A in Drake's Equation represents the number of new stars born in our galaxy in an average year. By studying the movements of star clusters around our galaxy, Newton's law of gravity shows that the mass of the galaxy is about 1,000 billion times that of the sun. About ten per cent of this is in the form of stars, the rest being dark matter such as dust and debris. Thus – roughly speaking – there are about 100 billion stars in our galaxy. Using the theory of how stars grow old and die, astronomers have estimated that the galaxy is about ten billion years old. So, on average, about ten stars have been born in the galaxy each year. We can therefore set factor A at 10.

Intelligent life is unlikely to live on a star, however – even the surface of a relatively cool star like our own sun is around 6,000°C, enough to vaporise any conceivable biological material. We need much cooler places – planets. Symbol B reflects this demand for ETI: it is the fraction of stars, such as the sun, which have planets in orbit around them. But to come up with a decent estimate of B requires a reliable theory of how planetary systems emerge in the first place – a question which, as we saw earlier in this chapter, is far from settled. For the sake of argument, let us suppose that planetary systems are formed by Woolfson and Dormand's capture process. In 1979, Woolfson estimated that about one sunlike star in 100,000 would gain a planetary system by this process. This means that B is about 1/100,000.

Is there any way of checking this? Planet-hunting beyond our solar system is a very specialised and delicate business, and few have tried it. In 1963, Professor Peter van de Kamp, then director of the Sproul Observatory in America, claimed to have found evidence of a planet orbiting around a faint star about six light years from Earth. His discovery was based on an analysis of over 2,000 photographs, dating back to 1916, of Barnard's Star. Barnard's Star appeared to be wobbling across the sky rather like a burly man waltzing across the ballroom floor with a diminutive partner. The implication was that

Barnard's Star had a Jupiter-sized planet in orbit around it. Evidence for a second, rather smaller, planet also emerged some years later. But analysis of van de Kamp's data and new observations carried out by American astronomers during the 1980s have cast grave doubt on van de Kamp's claims.

Rather more convincing evidence that planets may not be all that rare emerged in 1984, when a team from Arizona University published photographic images of a possible planetary system around the star Beta Pictoris, 53 light years from the Earth (a light year being the distance travelled by a beam of light in one year – about 10^{13} km). This is practically on our cosmic doorstep – our galaxy is about 100,000 light years across. So it seems reasonably likely that far more than one star in 100,000 has a planetary system. Even critics of the concept of ETI are willing to accept a figure of one star in ten possessing a planetary system. So let us settle on a value for B of 0.1.

Clearly, however, not all planets in a given planetary system are going to be suitable for intelligent life. This much is obvious from what we know about our own solar system. Venus is the same size as the Earth, and only a little closer to the sun, yet it could not support anything like human life. Drake built this fact into his equation with the factor C. Some planets, like Mercury in our own solar system, will be ruled out as being too close to the star at the centre of the system. Heat and other forms of radiation will break up the precursors of self-replicating molecules such as DNA and RNA, which are the hallmark of all (known) life. Other planets, like Jupiter and those beyond, will be too cold to sustain life-giving chemical reactions. Thus there will be a 'zone of hospitality' around a star where temperatures will be neither insufferably hot nor bitterly cold. We know that at least one of the sun's family of planets lies in such a zone – let us therefore set C as being equal to about 1 in 10, or 0.1.

But on what proportion of suitable planets does life *actually* emerge? With this question, we move from the 'vaguely reasonable' type of arguments to those based on the logic of 'in the absence of anything else to go on'. The question of what proportion of suitable planets give rise to life of some form is built into Drake's Equation in the form of the factor D. Many scientists think that simply too little is known about the chemistry underlying the origin of life to estimate D with any certainty at all. Others point to the relatively fast emergence of intelligent life on Earth – over the last three billion years or so – to argue that the process of life creation is not so difficult. So, in the absence of anything better to go on, let us adopt the ETI enthusiasts' estimate that between one in ten and every suitable planet develops life. Thus D is equal to 0.1–1; let's say 0.5 to be definite.

If you have not been very impressed by the way these estimates have been made so far, you're about to be even less impressed. Factors E, F and G are all but impossible even to guesstimate, so huge is our lack of knowledge. Factor E is the fraction of planets which develop *intelligent* life upon them and F the fraction of such planets that produce signs of their existence that we might detect. Finally, G is the average lifespan of that civilisation before, say, it blows itself up or chokes on its own pollution.

Guesstimates for these last three factors put forward over the years are more a reflection of the predilections of their originators than of hard fact. One could argue in a Darwinian sort of way that natural selection – 'the survival of the fittest' – always leads eventually to the emergence of intelligence. This would mean that E is close to 1 – all planets with life end up inhabited by intelligent life. One might also expect that F is close to 1 also – that is, a fair proportion of intelligent life forms ends up developing technology that could enable us to know of their existence. Not all, however: apes are clearly reasonably intelligent, but may always remain content not to communicate with the rest of the cosmos. The technology does not have to be very sophisticated – the development of radio may well be enough. Radio and TV broadcasts have been leaking out of our atmosphere and into deep space for decades. Travelling at the speed of light, these transmissions will have already told all ETs within 60 light years or so of the Earth with the right listening equipment that intelligent life exists somewhere else in the galaxy. So, bearing in mind the number of reasonably intelligent species on Earth who have yet to show signs of becoming radio hams, it seems reasonable to set F at 0.1 – that is, about one in every ten intelligent races has the ability to let us know of its existence.

And so we stagger at last to the final factor in Drake's Equation. This is rather different from the others, in that it is not a proportion lying between 0 and 1 but a *life span*: specifically, the life span of a technologically adept civilisation. We know from our own experience on Earth that such civilisations can exist for at least 50 years. But this may well be not just an underestimate of the life span of a civilisation, but a very substantial one: at least, we must hope so. Yet it may not be: perhaps all technologically advanced civilisations overreach themselves, polluting their world beyond repair, or destroying themselves by deliberate or accidental nuclear war – all spectres that haunt our society today. Perhaps intelligent life forms remain technologically advanced for only a relatively short amount of time before they realise that they must return to a simpler way of life. In short, the life span of an advanced civilisation is anyone's guess. The evidence from life on Earth is that it is unlikely to be less than 200 years. According to Carl Sagan of Cornell, an ET enthusiast, it may be as long as ten million years.

Clearly, the choice of life span has a profound effect on the outcome of Drake's Equation and its estimate of the number of advanced civilisations now existing in our galaxy. But after so much guesstimating, let us now work out this number. To make the calculation as interesting as possible, let us use the most optimistic of the various guesstimates we have made. We then have all we need to guesstimate the number of other civilisations in the galaxy. Simply multiply together all the numbers from A to G. This gives $0.1 \times 0.1 \times 0.1 \times 0.5 \times 1 \times 0.1 \times 10,000,000$. Thus there may be as many as 500 other civilisations in our galaxy.

This may seem really quite a large number – until one takes into account the vast size of the galaxy. Presuming that these civilisations are spread evenly through the main disk of our galaxy, one can show that each of these 500 civilisations would be separated from one another by about 4,000 light years – and this assumes the most optimistic values in Drake's Equation. Small wonder that no one has yet responded to our radio and TV broadcasts.

The search for extraterrestrial life

It is entirely possible that we are indeed alone in our galaxy. But the chance that there is even one other civilisation within a reasonable distance of us has been enough to stimulate some scientists to scour the heavens for evidence. The Search for Extra-Terrestrial Intelligence (SETI) programme really started in 1959, when two American astrophysicists wrote to the influential science journal *Nature* to describe how such a search might be set up. The idea is simple enough in principle. Just build a sensitive 'ear' capable of hearing the transmissions coming from the other civilisations, point it at the sky and wait. Radio receivers are the obvious choice for the 'ear' needed to pick up, say, the radio transmissions that leak from an inhabited planet. But such receivers must be extremely sensitive if they are to pick up the vastly diluted strength of the signals from a distant planet. However, they already exist: the largest radio 'ear' yet built is the 305-metre diameter radio telescope at Arecibo in Puerto Rico. This is capable of picking up any such transmissions from the planets of nearby stars.

The other unknown in the SETI question is exactly which frequency the extraterrestrials will be broadcasting on. High-speed digital processors now exist which can automatically scan huge frequency ranges, searching for any signs of intelligent communication. SETI technology has thus now reached the stage where a concerted effort to search for intelligent life elsewhere in the galaxy can begin.

In 1992 NASA, the US space agency, will begin a $100 million search programme to do just that. The search will be split into two directions. One team will use the 34-metre radio telescopes of NASA's deep space network (DSN) to scan the whole sky over a broad range of frequencies up to 25,000 megahertz (MHz), searching for evidence of powerful interstellar beacons planted by ETs. One particular frequency range will be monitored with particular care: that around 1421 MHz. This is the so-called 21-centimetre line, of great significance in astrophysics because it is associated with the presence of hydrogen – the most common element in the galaxy. Many SETI experts think that ETs may choose this universally important frequency so that any other intelligent beings would have a clue of which frequency to concentrate their search on.

This all-sky search will take five years. It will be 10,000 times more extensive than any previously carried out, and 300 times more sensitive.

The second team will concentrate their efforts on the hundreds of stars within 80 light years of Earth, searching for signs of radio 'leakage' from inhabited planets. Again, a broad range of frequencies will be covered, including the 21-centimetre line. However, the sensitivity needed will be much higher, demanding the use of the huge Arecibo radio telescope and the 64-metre telescopes of NASA's DSN. This search will take about three years.

Will the search be successful? 'Possibly, but not probably' seems a reasonable assessment. Looking for leakage of broadcasts may founder because the ETIs have switched to more sophisticated technology. In the early days of radio and television, powerful transmitters were the only way of getting broadcasts around the world. We are now seeing the advent of satellite broadcasting, which gives the same coverage without the need for such power. The amount of leakage from such networks is much smaller – perhaps undetectable by the NASA technology.

The search for transmissions from deliberately planted beacons assumes that ETIs actually want to be discovered. But perhaps they know something we don't, and have decided that attracting attention is not a very bright idea. It is usually assumed that advanced civilisations are benevolent, having risen above mindless violence against lesser races. But what if the ETIs have made such a mess of their planet that they are looking to hijack another? They may turn out to have the benevolence of a crew of a sinking ship scrambling for the nearest lifeboat.

Some scientists expressed concern about giving an open invitation to visit Earth when the American probe Pioneer 10, the first to leave the solar system, was fitted with a map and other data to tell anyone and anything that bumped into it where it had come from.

Many scientists, however, think that the NASA search will not succeed for a rather simpler reason than any of this: ETIs do not exist. While many of these sceptics base their belief on nothing more sound than a lack of imagination, some have come up with rather more detailed objections. Dr Frank Tipler of Tulane University in New Orleans argues that if ETIs exist, they should have made their presence felt already. He proposes that an advanced civilisation will want to explore the galaxy using technology that minimises the cost of the information obtained. Tipler therefore suggests that the ETIs would soon realise that the best approach is to use probes which can exploit the resources of the planets they discover, repairing and replicating themselves – in other words, 'robot pioneers'. The technical term for such beings (of which we are an example) is Von Neumann machines, after the Hungarian-American mathematician Johnny Von Neumann who developed the mathematics behind them. In Tipler's 1980 paper on the implications of these machines for the ETI debate, he calculates, using the Drake Equation, that 'If extraterrestrial intelligent beings exist, then their spaceships must already be present in our solar system'. Up to this point, Tipler's argument seems reasonably convincing, at least by the pretty poor standards of the ETI debate in general. But then he claims that since we don't see any Von Neumann machines, ETIs do not exist. Frankly, I think this line of reasoning has more holes in it than a hundredweight of Gruyère cheese. Let us take just three objections, although there are many more. It is by no means certain that ETIs would choose to use Von Neumann machines to explore the galaxy. Even the relatively barbarous race that now inhabits the Earth is concerned about the amount of junk in space – the idea of polluting the entire galaxy with countless thousands of self-replicating robots would surely be repellent to an advanced race.

But suppose that the ETIs are hell-bent on using Von Neumann machines. Then is it inconceivable that, like a bird-watcher in his camouflaged hide, the machines have been designed to observe undetected, lest they disturb – possibly even panic – the civilisation under observation? Perhaps the Von Neumann machines are already here.

Finally I suspect that Tipler's statement that, if ETIs exist their spaceships must already be here, may have triggered off an immediate reaction in some readers of 'Well, they are already here! They're called UFOs.' The ETI debate is notorious in scientific circles for its substantial amount of conjecture and its unconscious assumptions. But even ETI enthusiasts like Sagan, while attacking Tipler on the details of his arguments, will never call on UFO sightings to back their case. The reason is simple – if a scientist starts to make references to UFOs he or she incurs instant derision from his peers. It is professional suicide. It also virtually guarantees the scientist a flood of letters and telephone calls from every fruitcake in the land. Thus no scientist can call on what many ordinary people believe is *prima facie* evidence for the existence of ETIs to support or contradict a pro- or anti-ETI argument. I have no strong feelings either way about UFOs, but I can't help thinking that by simply dismissing them from all scientific debate, some important constraints on ETI theories are being missed.

But just before we pull back from the abyss of non-science, I cannot resist one last, rather disturbing, reason why NASA's search should fail. It is that Von Neumann machines, designed by a race of psychotic ETIs, are indeed abroad in the galaxy. These deadly probes seek out radio leakage from inhabited planets and then move in for the kill. As Dr Glen Brin, an ETI-debate expert at the California Space Institute at San Diego, points out: 'The frightening thing about "deadly probes" is that it is consistent with all of the facts . . . We would not have detected extraterrestrial radio traffic – nor would any [ETIs] have ever settled on Earth – because all were killed shortly after discovering radio.'

There are, then, many reasons why we might expect the NASA search to fail. We may not have heard from anyone yet simply because the ETIs have chosen not to communicate. As Brin poetically puts it: 'It might turn out that the Great Silence is like a child's nursery, wherein adults speak softly, lest they disturb the infant's extravagant and colourful time of dreaming.'

But the key point in the debate about whether or not extra-terrestrial intelligence exists was most neatly made by Carl Sagan: 'Absence of evidence is not evidence of absence.'

The vast majority of scientists view the SETI debate with the utmost scepticism. This sets them apart from the average person, who is inclined to believe that life may well exist somewhere else and that, indeed, it may even have already visited us. This conflict of views may well have contributed to a widespread perception of scientists as being tedious sceptics all too ready to pour cold water on any idea deemed too exciting and comprehensible. Yet at the very heart of science lies a theory whose predictions are so strange and magical that it strains the human imagination to breaking point: quantum theory.

4 Enigmas at the Root of All Things

O F THE MANY CLICHÉS ABOUT SCIENCE, arguably the most fatuous is that 'Science is just organised common sense'. Whoever coined it clearly knew nothing about quantum theory – the rules that underpin the behaviour of all matter in the universe.

A lot of what follows is going to seem very strange indeed. You should not imagine that this is because of any intellectual failing on your part; the late Richard Feynman, Nobel prizewinner and one of the most brilliant exponents of quantum theory, once remarked: 'You never understand it, you just get used to it.'

But before we plunge into the quantum jungle, one thing should be made clear. Quantum theory is not some quasi-philosophical mumbo jumbo. It is a scientific theory, indeed the most successful ever conceived. Its ideas can and have been turned into unequivocal predictions which have been tested in the laboratory. And every time quantum theory has been put to the test it has, for all its strangeness, turned out to be right.

It is also crucial to understand that when we meet – as we shall shortly – the multiple universes and undead cats that emerge in the search for the foundations of this almost magical theory, we are still dealing with ideas that underpin the workings of such practical things as transistor radios and body scanners.

If nothing else, by the end of this chapter you will know the truth of J.B.S. Haldane's contentions that the universe 'is not only stranger than we imagine, it is stranger than we *can* imagine'.

The birth of a strange theory

At the turn of the last century, an air of something close to complacency was beginning to permeate science. It seemed that a combination of mathematics and certain astonishingly powerful instruments had given scientists the ability to take an Olympian view of the universe, and know more or less everything about it.

It was not empty arrogance. During the nineteenth century, science had achieved some truly impressive successes. In 1846, using the law of gravity discovered by Sir Isaac Newton almost two hundred years earlier, mathematicians from England and France predicted the existence of a previously undiscovered planet – Neptune – thereby increasing the size of the known solar system by 50 per cent, using just pen and paper.

With the invention of the spectroscope, an instrument for splitting white light into its rainbow of constituents, scientists had achieved something almost miraculous: a way of discovering the chemical composition of the sun and stars without ever leaving Earth.

The successes were not of just an academic nature. Victorian engineers were taking the discoveries of scientists such as Michael Faraday and building machines that were changing the course of history.

It was thus somewhat inevitable that some scientists were beginning to think that maybe the end was in sight – that a single, albeit large, textbook would be able to contain all the laws necessary to understand every phenomenon from the origin of man to the movement of the distant stars.

A century later we are still waiting for that book to emerge. Curiously enough, the reason we are still waiting is rooted in discoveries made in a field of science that seems rather unlikely a birthplace for a revolution: the theory of heat.

In the mid-nineteenth century, a German physicist named Gustav Kirchhoff was developing a theory to explain how heat is radiated from one place to another. To make his research a little easier, Kirchhoff had invented the notion of an object that absorbs all the heat and light that falls on to it. It is called, somewhat unimaginatively, a 'black body'. Inventing such idealised objects is common practice in physics. It is often the only way of getting started on an explanation of an otherwise hideously complex real-life problem.

Having invented the black body, Kirchhoff and others set about studying its properties, both theoretically and experimentally. To mimic them, they heated containers with blackened insides, and studied the 'black body radiation'* that emerged from them via a small hole in one side.

The aim of the experiments, roughly speaking, was to study how this radiation changed when the temperature of the black body was altered. Armed with this knowledge, it would then be possible to make comparisons between black bodies and much more interesting hot objects – like the sun – and see if the same theory seemed to apply to both. If it did, then the black body idealisation might lead to some unexpected discoveries about the sun and stars.

By the late 1890s, some progress had been made. Wilhelm Wien, another German physicist, had studied how the medium and high frequency radiation coming out of a black body changed with temperature. By some judicious algebraic juggling, Wien also found a formula that gave the amount of energy as a function of both wavelength and temperature for the range of frequencies he had studied. Meanwhile, in Britain, Lord Rayleigh had worked out another formula which did the same for the low-frequency black body radiation.

Having two formulas that do the same job in two different circumstances may suit an engineer, but for physicists it is deeply unsatisfactory. Wien's formula was useless at low frequencies, and the Rayleigh was useless at high frequencies. Yet surely a single, overarching law applied to the black body radiation? To a physicist, this belief is based on a deep-seated conviction in elegance. Two laws for the same phenomenon is, to a physicist, appallingly inelegant.

It was obvious that what had to be done was to find a formula that boiled down to Wien's rule when applied to high frequency radiation, and to Rayleigh and Jean's rule when applied to low. Such a single law would, presumably, also give the right answer for all the frequencies in between.

In October 1900, Max Planck, a 42-year-old physics professor at Berlin University, was working on the task of finding this one, unifying law of black body radiation. And, in the early evening of Sunday 7 October, he found it. Although he was yet to realise it, he was about to trigger one of the most profound revolutions in the history of science.

---

* The word 'radiation' is often used simply in relation to nuclear processes – i.e. radiation in the form of fast-moving particles and gamma rays. In this book, the word is used in its much wider context of describing those things that are radiated, e.g. light radiation from a lamp, heat radiation from a fire.

The problem for Planck was that he could not be content simply to juggle some symbols around to give a formula that did the trick he required of them. He could not leave his new-found formula as just, in his own words, 'a luck intuition'. He wanted to find out *why* the formula was as it was – and what physics lay behind it.

His search for an explanation led him to take the first step across the boundary of the safe, familiar physics of Sir Isaac Newton and into the bizarre, disturbing realm of the quantum. Planck discovered, to his astonishment, that to arrive at his formula, he had to give up the idea that the black body absorbs light and heat *steadily*. He was forced to assume that it could absorb radiation only in discrete packets. These he called 'quanta', from the Latin word for 'how much'. Planck knew immediately that he had uncovered something very deep about Nature. He took his little boy for a walk and told him: 'I have had a conception today as revolutionary and as great as the kind of thought Newton had.'

Planck went on to produce a law that related the energy, E, contained in one of these radiation energy packets to the frequency, f, of the radiation it carried. He showed that one was related to the other by the formula $E = hf$, where h is a constant, now called 'Planck's constant'.

This constant, h, is a very small quantity. The implication of this, roughly speaking, is that it is only when one is dealing with very small things such as atoms that one has to worry about energy coming in packets. When trying to explain the behaviour of most everyday objects, therefore, one would not notice quantum effects. This explains why no one had – until now.

All very neat. But for many of Planck's contemporaries – even for Planck – there remained a feeling that quanta might really be little more than a mathematical trick that explained another mathematical trick that explained something about black body radiation. For five years, Planck's ideas languished for want of hard evidence that quanta were anything more than figments of a physicist's imagination.

Then in 1905, a struggling nobody named Albert Einstein, working in the Berne Patent Office, published a paper that would provide that evidence. Once again, the breakthrough came from work in a seemingly unrelated field: the effect of shining light on metals.

In 1887, the German physicist Heinrich Hertz made the surprising discovery that if he shone high-frequency light on to certain metals such as silver, he could produce electricity. These electrical effects turned out to be caused by electrons which came flying off the surfaces of the illuminated metals.

This so-called photoelectric effect is very familiar today through its myriad applications. By shining a beam of light across a doorway on to a photoelectric cell, for instance, one can create a burglar alarm. When the light beam is broken, a sudden drop in the number of electrons flowing from the photoelectric metal into a detector triggers the bell.

But the photoelectric effect proved to be deeply enigmatic. For example, one would have expected that to get electrons to come flying off the metal faster, one should use a brighter light. But no: experiments found that, although boosting the intensity increased the *number* of electrons that came off, their individual energies were unchanged. To boost their speed, one had to increase the *frequency* of the light. There were other mysteries. No electrons emerged at all unless a certain frequency – rather than brightness – of light was shone on the metal. But then the electrons emerged immediately, even if very faint light was used.

Young Einstein makes his debut

At this point, the 26-year-old Einstein enters the picture. He had been worried for some time about a simple, yet profound question: should theories based on waves, which are smooth, continuous things, be used to solve problems involving discrete objects such as atoms and molecules?

Einstein found an answer by studying the implications of Kirchhoff's black body concept. Like Planck, he found that it led to a startling, and deeply enigmatic, conclusion. Einstein discovered that radiation such as heat and light – until then regarded as wavelike flows of energy – could sometimes be regarded as made up of individual packets of energy.

One should pause for a moment to ponder the revolutionary nature of what Einstein had uncovered: a single phenomenon, with two utterly different characteristics, one smooth and continuous, the other discrete, or 'quantised'.

Einstein put his quanta to work explaining the photoelectric effect. The most baffling part of this phenomenon, the dependence not on intensity but on frequency, was now instantly explained. If light is thought of as a shower of individual packets of energy, increasing the intensity would obviously make no difference to the number of electrons leaping from the surface of the metal. It would increase merely the *number* of quanta hitting the metal, not their ability to knock electrons out of the surface. Think of beating a carpet: if you hit it lightly with your hand once or a thousand times, you won't get rid of the ingrained dirt. What you have got to do is hit it really hard – that is, put in a lot of energy. And, according to Einstein (and Planck), the energy contained in a quantum of radiation depends solely on its frequency.

Einstein's explanation was neat – but again, was it anything more than that? Was it just a mathematical trick, or a reflection of a genuine duality in the fundamental nature of matter? In 1909 Einstein pointed out something that strongly suggested the latter. He showed that the formula for *changes* in the energy coming from a black body fell naturally into two pieces. One half had the form one expected if radiation was wavelike, the other half the form expected if radiation came in packets. So now Einstein could call on both theoretical and laboratory-based arguments to back his contention that radiation could be regarded as coming in packets of energy. People started to take notice, especially of the explanations of the otherwise baffling experiments with black bodies and photoelectric metals.

It soon emerged that light and heat radiation were not alone in showing signs of curious quantum properties. In 1911, Ernest Rutherford at the Cavendish laboratory in Cambridge University had shown that atoms making up all the objects around us could be pictured as a miniature solar system. Bound together by the electro-static force (the same force which enables a comb to pick up bits of paper after being brushed through one's hair), negatively charged electrons are in orbit, like planets, about a positively charged 'sun' – the nucleus.

It was a comforting picture, but one at whose heart lay a profound problem. Physicists already knew that charged objects lose energy when they change speed or direction. An electron circling around a nucleus is in just such a situation. So the electrons should lose energy as they orbit the nucleus, spiralling in and hitting it in about 10^{-17} seconds. In other words, no sooner has an atom been made than it collapses. No atoms means no materials – and no intelligent life like Rutherford.

It was a desperate state of affairs, calling for desperate measures. Once again, it fell to a Young Turk to come to the rescue, brandishing quantum ideas. Niels Bohr, a 27-year-old Danish physicist, put forward an idea that, if correct, would explain why electrons do not rapidly collapse on to the nucleus. He proposed that an electron cannot orbit the nucleus at just any distance it pleases. It could, said Bohr, sit only in certain specific orbits, whose sizes are some exact multiple of a fundamental orbit. Bohr showed that this proposal leads to the conclusion that an electron cannot have any energy it likes. It can have only an energy that is an exact multiple of a certain, fundamental packet – yes, quantum – of energy whose value he could calculate. For each allowed orbit there is a corresponding energy, and when the electron is in one of these stable orbits, said Bohr, it does not lose any energy, contrary to the usual rule. Thus the electron will not spiral into the nucleus.

It is interesting to ponder on just what sort of reception this daring but rather *ad hoc* proposal would get from referees of today's science journals. Fortunately for Bohr, he was able to provide backing for his proposal by showing something quite extraordinary: that Kirchhoff, the inventor of the black body, had actually seen evidence for the quantum nature of the atom as far back as the 1850s.

When not working on the black body, Kirchhoff had been studying the effects of heating chemicals in flames. Viewed through a spectroscope, which splits up light into its constituent frequencies, or 'spectrum', the incandescent chemicals turned out to have spectra consisting of a series of individual, brightly coloured lines. For example, hydrogen has three vivid lines of red, blue-green and blue in its visible spectrum.

Using maths that would not trouble a sixth-former, Bohr was able to show that the arrangement of these lines is a direct result of the quantum nature of the atom. Although electrons in stable orbits do not give off energy, they can be kicked into higher stable orbits, or even out of the atom altogether, by putting energy into the atom – for example, by heating the material. Of course, the electron must give back this energy as the atom cools, and it emerges as light of a frequency – and hence colour – given by Planck's simple law $E = hf$.

But, according to Bohr, the electron's energy is quantised, and so the light given out cannot be of any frequency. The frequency is then related by Planck's simple law (derived, remember, from the theory of heat, of all things) to the energy differences between the orbits between which the electron jumps. So for each 'quantum jump' down the electron makes, the light that emerges will appear as a sharp spectral line of a certain, very specific frequency – just as Kirchhoff saw all those years earlier. Bohr was even able to work out what colours should be seen for heated hydrogen – the simplest of all atoms, consisting of just a single electron orbiting a single proton. In doing so, he found an explanation for a hitherto mysterious 'law' that connected these colours to simple numbers, discovered by an amateur Swiss scientist named Johann Balmer way back in 1885.

Bohr then pushed his theory further, to give an account of the light from helium, the second simplest atom. Despite its relative simplicity, he succeeded in showing that puzzling observations of a star in the constellation Puppis were the result of hot helium gas surrounding the star.

Without such a brilliant display of the power of his new theory, Bohr's ideas would never have been countenanced by the rest of the scientific community. But even so, many scientists still looked down their noses at all this jiggery-pokery. Einstein may have explained the photoelectric effect by thinking of light as being made up of showers of these pesky packets of energy, but everyone 'knew' that light is really made up of waves. After all, had not incontrovertible proof for this been obtained a century earlier?

This 'proof' that light really is wavelike was indeed impressive. Thomas Young, an English physician and physicist, discovered that when light was allowed to pass through two slits on to a screen, the two beams produced would 'interfere' with one another, causing a pattern of alternating light and dark on the screen. This curious pattern was easily explained with the wavelike nature of light. When the waves making up the two beams interfered with each other, at some places the peaks in each beam would coincide, producing a really bright region, while elsewhere a peak in one wave-train would coincide with a trough in the other, cancelling each other out.

Young's demonstration initially won him only the opprobrium of other British scientists – essentially because he had provided evidence that the great Isaac Newton had been wrong. Newton had been a powerful advocate for the idea that light was essentially a stream of particles. But Young's experiment defied explanation on this basis, and eventually his experiment was seen as the triumph for the wavelike theory.

And here we reach the first paradox of quantum theory. Einstein and others have produced incontrovertible evidence that light behaves like a shower of quanta. Young has shown incontrovertibly that light behaves like a wave. Must we therefore think the unthinkable – that Einstein and Young were *both* right? Can light be both wave and particle *at the same time*?

The clinching evidence for this revolutionary new view of the nature of all matter and radiation came in 1923. An American physicist, Arthur Compton, found that when x-rays were shone on to thin metal foils, some of the radiation went straight through and some of it was reflected. But, most interesting of all, the x-rays that were scattered off the foil had a different, lower frequency.

The simplest explanation of this phenomenon, which became known as the Compton effect, was that x-rays, like light, can behave both as packets of radiation and as waves. What was happening in Compton's experiments was that quanta of x-ray energy were smacking into electrons making up the atoms in the foil, and passing on some of their energy to the electrons, just like one snooker ball hitting another. The quanta, having lost some of their energy, must, by Planck's law $E = hf$, have a reduced frequency as well – just as observed. Compton found that, by considering x-rays as packets of energy rather than as waves, he could account for what he had seen in every detail.

Compton's experiment had finally established the dual nature of all radiation – what has become known as 'wave-particle duality'. There was no real paradox inherent in the simultaneous validity of both Young's experiments and, say, the photoelectric effect: they were experiments which simply showed different aspects of the same thing.

The prince and the particles

The first battle of the quantum revolution had still one last blast to fire, and the man who lit the fuse was a 32-year-old French aristocrat, Prince Louis de Broglie. His family were no strangers to revolution – his great-great-grandfather had served the royal court of France and been executed at the guillotine. De Broglie had decided to make science his life's work after sitting in on an international conference on physics in Brussels in 1911. He took a degree in science and then went on to write a doctoral thesis. But this was to be no ordinary thesis. In it, de Broglie put forward what seems like a simple extension of quantum ideas. As so often with quantum theory, however, nothing is quite what it seems.

De Broglie pointed out that the photoelectric effect and the Compton effect showed that radiation – previously thought of as always wavelike – could sometimes behave as if it were made up of particles. So, asked de Broglie, could not particles like electrons and protons have some wavelike features too?

De Broglie worked out the consequences of this simple idea, arriving at a formula that enabled the 'wavelength' associated with a particle to be calculated. What was needed now was some experimental evidence for matter behaving like waves. It emerged in 1927. The American physicist Clinton Davisson found that it was possible to form a kind of interference pattern by beaming electrons down on to a crystal of nickel. It seemed that the regular arrangement of atoms making up the crystal were acting like the slits in Thomas Young's experiment. Clearly, the electrons had something wavelike about them which enabled them to show this very wavelike behaviour.

The prince had been right. All matter, even electrons, has wavelike properties. Some experiments reveal the billiard-ball nature of particles, others their wavelike properties.

And at this point, in the late 1920s, we reach a watershed in the development of quantum theory. Let us take stock. Experimental evidence had shown beyond reasonable doubt that what started out as a mathematical trick really did apply to the real world. Radiation and matter are not just waves, or microscopic billiard balls. They are some kind of combination of both – 'wavicles', if one likes.

The notion of this wave-particle duality is hardly obvious, but it can just about be stomached if we accept that perhaps in the past we have been too keen to describe all of Nature in terms of everyday things such as waves or billiard balls. Perhaps Nature is rather more complex.

But, as we shall now discover, de Broglie's idea of *particles* having wavelike properties is more than just the flipside of the idea that *radiation* is quantised. For it is here that quantum theory takes us into territory far beyond our everyday experience. Like pilots flying into dense fog and switching to instruments, physicists trying to understand the next developments in quantum theory found themselves having to put their faith not in physical intuition, but in mathematics.

What they discovered about the realm of the quantum proved too shocking for Einstein. They challenged his fundamental belief in the way the universe worked. A latter-day Danton, Einstein was consumed by the very revolution he had started.

Ironically, Einstein was one of the earliest supporters of de Broglie's idea. It did, after all, explain a number of otherwise puzzling features about this new quantum theory. For a start, de Broglie's central idea that every particle had a wave associated with it – he called them 'pilot waves' – provided a solid (or at least, *somewhat* more solid) foundation for Bohr's model of the atom and his explanation of why it doesn't collapse. Recall that Bohr had shown that this embarrassing problem could be avoided if electrons couldn't just orbit anywhere they liked. He had come up with a restriction on the types of orbit allowed, and the consequences of that restriction accorded very nicely with the experiments (up to a point).

But Bohr's restriction seemed to have been plucked out of the air. De Broglie seemed able to give an explanation for it: the allowed orbits were those where the pilot wave associated with an electron fitted nicely, with no overlap.

So far, so clever. But could these pilot waves account for anything else? What was needed was a recipe – a formula – for working out how such waves behave when they are doing things more interesting than merely sitting in simple atoms.

In 1925, a 38-year-old Austrian professor at the University of Stuttgart, named Erwin Schrödinger, was leafing through a paper by Einstein when he saw a footnote describing de Broglie's work. Schrödinger decided to set about trying to find such a formula. Now called Schrödinger's equation, it has become one of the most famous in all science, ranking with Einstein's $E = mc^2$, though it is decidedly more complex.

The formula is a judicious mix of de Broglie's ideas, some standard mathematical stuff for describing any wave and a bit of simple mechanics. The whole thing is held together by the requirement that it give results consistent with what little experimental data were then available about these pilot waves: specifically, the results of Davisson's nickel crystal experiment. These all seem eminently sensible things to demand of the formula. Yet the end result of Schrödinger's mathematical efforts were anything but common-sensical. The 'pilot

How quantum theory has modified our view of the atom

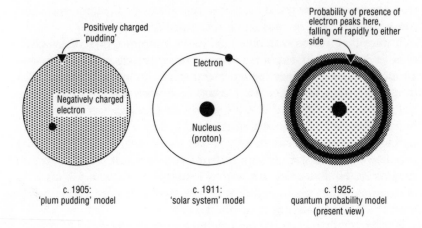

waves' turned out not to be like ordinary waves at all. They are, in a mathematical sense, 'unreal'.

Schrödinger was deeply disturbed by this curious turn of events. 'Pilot waves' had been invented to do something concrete – explain the existence and workings of every atom in the universe. But Schrödinger's equation was telling him that pilot waves are really very abstract creatures. He decided that 'pilot waves' conjured up too quotidian an image; to reflect their more abstract nature, Schrödinger dubbed them 'wave functions'.

Whatever they were, the wave functions seemed to do a fine job of accounting for the past successes of quantum theory. Plugging the characteristics of a hydrogen atom into the equation immediately led to all the results obtained by Bohr's model. The packetlike nature – quantisation – of energy popped out automatically. The equations could even deal with complex problems posed by atoms with more than one electron, which Bohr's theory really struggled with. Many physicists were suitably impressed by Schrödinger's formula; among them was Einstein. 'I am convinced that you have made a decisive advance,' he wrote to Schrödinger.

The end of certainty

One reason for the plaudits was psychological: Schrödinger's equation provided a comforting feeling of familiarity in an area of physics that seemed to be showing signs of becoming too esoteric for its own good. A completely new and intensely mathematical way of looking at quantum theory was emerging, and its prime movers were the 25-year-old German physicist Werner Heisenberg and his professor and mentor Max Born. Based on the manipulation of mathematical objects called matrices, the new version of quantum theory was the apotheosis of abstraction. Particles as billiard balls or waves? Naive and illusory, said Heisenberg. Forget about trying to form any mental pictures. With quantum theory, only what we measure in the laboratory is known for certain. Using this philosophy and the basic rules of mathematics, we shall derive everything, he proclaimed.

Einstein thought it all very clever, but didn't care for it one bit. 'It's a real witches' calculus . . . most ingenious, and adequately protected by its great complexity against being proved wrong,' he told a friend. It was becoming clear, however, that the old ways of thinking about the atom weren't going to last much longer. Schrödinger's equation seemed the only solid, familiar ground in the bizarre and abstract realm of the atom.

And then Max Born turned even this into a quicksand.

Just a few months after Schrödinger had published his famous equation, Born put forward an interpretation of the curious wave functions whose behaviour the equation described. There is a simple reason why these waves seem rather unreal, said Born. It is because they are nothing more than mathematical devices. However, all is not lost. By using a simple rule, said Born, it is possible to get some hard information out of them concerning the particle associated with the waves.

Yet this 'hard information' turned out to be distinctly woolly. One perfectly reasonable thing to ask of a theory about electrons is where an electron is at any particular time. But according to Born's idea of what quantum theory was all about, Schrödinger's equation could not say *exactly* where the electron is – it can only give the *probability* of the electron being in a certain place. Suddenly, Schrödinger's equation no longer looks like a little oasis of sanity in a vast ocean of quantum weirdness. Ask it a straight question and it can't give you a straight answer. It emerged that Schrödinger's 'eminently reasonable' equation was just another way of writing Heisenberg's appallingly abstract 'matrix mechanics'. There was to be no hiding place for those who could not tolerate the quantum revolution.

Fittingly, it was Heisenberg who delivered the *coup de grâce* to classical physics, as pre-quantum physics has now become known. In 1927, in one of the most famous scientific papers ever written, he exploded one of the foundations of science. He showed that the new quantum theory implies that there are fundamental limits to our ability to know everything about the world. This discovery has now become enshrined as Heisenberg's famous 'uncertainty principle'.

Specifically, the principle states that it is impossible precisely to measure both the motion and position of, say, an electron. If you pin down exactly where the electron is, you cannot say with equal precision what its velocity is. Conversely, if you measure the velocity of an electron very accurately, any attempt to locate the electron exactly is doomed to failure.

The fundamental limit to knowledge, it turns out from the mathematics, is dictated by the size of Planck's constant. For those readers with a bit of maths, if x is the error in the observation of position or energy, and y is the error in the momentum or characteristic time, then the uncertainty principle states that xy is at least as great as Planck's constant, h. If this constant were zero, there would be no problem; we would be back in the world of classical physics and its apparently limitlessly accurate experiments. Though small, Planck's constant is not, however, zero. And in the world of the atom, it is too big to be ignored.

The limits to knowledge set by the uncertainty principle are not, claimed the revolutionaries, merely empirical. They are absolute and cannot be circumvented by using, say, more accurate measuring instruments. They are a fundamental property of the universe we inhabit.

Clash of the giants

For Einstein, Heisenberg's work was the last straw. A classical physicist in the mould of Newton, he could never accept that there are limits to knowledge, that the world is, at root, essentially and ineluctably unknowable. Having started the quantum revolution and led it for many years, he now became its most redoubtable critic.

His opponent was Niels Bohr, the Young Turk who had now emerged as leader of the quantum revolution. In 1927, Bohr issued what amounted to the manifesto for the revolution. He brought together all the strands of the new theory into what he claimed was a consistent way of looking at the world – a philosophy of reality, no less – which could handle the implications of the quantum theory. He was forced to make amendments to the manifesto over the years, however, in response to the brilliant challenges of Einstein, who spent much of his later years trying to find ways around those aspects of quantum theory he found repellent.

The end result of this great debate between the giants of twentieth-century physics has become known as the 'Copenhagen interpretation' of quantum theory, after the city in which the Niels Bohr Institute of Physics was established. It remains the most widely accepted way of looking at quantum theory among scientists today, but it should be stressed that, despite what some might claim, it remains just that: an interpretation. There is powerful experimental evidence that it is an acceptable way of looking at the world, but it is most definitely *not* the only possible interpretation.

In his 'manifesto', Bohr first crystallised the idea of things like electrons behaving sometimes like waves (as in, say, the Davisson experiment) and sometimes like particles (as in collision and scattering experiments) in the concept of *complementarity*.

According to this, wavelike and particlelike behaviour are just complementary aspects of a quantum object's fundamental nature. They can be studied to one's heart's content by carrying out different experiments. According to Bohr, every experiment reveals something either about the wave nature, *or* about the particlelike nature – but no experiments will ever throw up any conflict between these two different aspects.

There is, according to the Copenhagen interpretation, an explanation for this – and it is as profound as it is startling. It is that the act of observation brings the things we are observing into existence. According to Bohr, the act of observation transforms the fuzzy, unpredictable denizens of the quantum world into the measurable, 'real' objects in our own world.

This is why we will never encounter an experiment that can give us information about both the wavelike and particlelike behaviour simultaneously. The mere act of carrying out the experiment dictates what type of behaviour – wavelike or particlelike, we shall find.

Can it be that the electron goes about its business of being a 'wavicle' – an entity with both types of behaviour – when we are not observing it? Not according to Bohr, who took a stridently positivist line on such questions. Physics can never answer questions about the nature of an electron not under observation. Getting knowledge about something inevitably means performing some observation on it. And that means disrupting its raw state.

How to create reality

How can the act of asking questions about an entity bring that entity into existence? To help those who are justifiably incredulous, American physicist Professor John Wheeler has devised an elegant analogy, drawn from everyday life, that shows how it is possible to create something simply by trying to find out about it.

Professor Wheeler's analogy is based around his attendance at a dinner party, where the after-dinner guests are playing the game of Twenty Questions. Professor Wheeler's turn comes round and he is sent out of the room while the others think up some suitably tricky target for his questions.

He is locked out of the room for an inordinately long time. When he is finally let back in, he sees that all the guests have smirks on their faces – they've clearly cooked up something clever. So the Prof starts his questioning, fully expecting something odd to happen. To start with, everything goes as normal. He asks 'Is it animal?' and they say 'No'; 'Is it mineral?' and they say 'Yes'. But after a while the responses are slower in coming – despite the fact that only simple yes/no answers are needed.

Battling on, the Prof suspects the target word may be 'cloud'. He reaches his final, twentieth question – 'Is it a cloud?' he asks. 'Yes!' comes the reply, and everyone roars with laughter. They then explain that, when he came into the room, *they had no word in mind at all*. They had decided to make the game more interesting for themselves by giving responses solely on the basis that each was consistent with all the preceding responses – hence the delay as the game progressed. Bit by bit, the questioning of the Prof led from nothing at all to the formation of a consistent answer – 'cloud'.

Had he asked a different set of questions, he would obviously have got a different end result. To this extent, he had control over the nature of the thing he was bringing into existence. Clearly, much also resided in the responses of the other guests. But the central point is this: the target of the questioning and the method of the questions are inextricably tied up with one another.

And so it is in the laboratory, according to Bohr. Experiments are nothing more than ways of asking questions of reality, and the way they are set up has a profound effect on what you will find from them. You cannot separate the experimental apparatus from the thing being studied.

The Copenhagen interpretation has one other central dogma: the absolute unavoidability of uncertainty. According to Bohr, no amount of ingenuity will ever enable us to determine with complete accuracy everything we want to know about the behaviour of a particle. The best we can do is to use things like Schrödinger's equation to give the *odds* that an electron is at a particular place, say, at a particular time.

Again this is hard to swallow and indeed Einstein never did accept it; he summed up his views on uncertainty and the probabilistic nature of the quantum world in his now famous quotation, 'God does not play dice with the universe'. He went on to attack Bohr's manifesto for making such assertions, believing that all this philosophical stuff was just a cover-up for a pernicious defeatism underlying quantum theory.

Einstein's arguments have a strong emotional appeal. In the ordinary world, uncertainty is something we are very familiar with. It is the result of not taking into account all the relevant factors involved in a situation. The reason one cannot tell *exactly* where a coin will land when you throw it is not because of some intrinsic indeterminancy, but because we simply haven't taken into account all the variables, such as air resistance, that affect the end result.

Einstein took a similarly common-sense view of the uncertainty in quantum theory. It is, Einstein believed, simply the result of not trying hard enough to discover what is really going on behind the scenes. According to this view, quantum theory is incomplete. There are hidden variables at work in the theory which would enable us to calculate *precisely* what particles are doing – if only we knew the values of these hidden variables.

During the later years of his life, Einstein thought up a number of ways of blowing holes in Bohr's manifesto, and proving that common sense could still prevail in the realm of the quantum. His most famous attack – the EPR thought experiment – was made in collaboration with physicists Boris Podolsky and Nathan Rosen while all three were at the famous Institute for Advanced Study in Princeton in 1935. Its target was Bohr's positivist claim that particles do not have definite properties until someone makes an *observation* of them.

The attack took the form of one of Einstein's famous 'thought experiments'. An imaginary picture of an experimental set-up, it is an amazingly powerful technique that gives not quantitative data but a way of detecting contradictions.

Picture a single molecule, made up of two particles, which explodes: one particle flies off to the left, the other to the right. Now, according to Bohr, until an observation is made of a particle, it cannot be said to have such properties as velocity or position.

But wait: what about Newton's famous law that every action generates an equal and opposite reaction? This means that the two particles fly out in opposite directions but with the same speed. So measuring the speed of one of the particles allows us to know, with exactly the same precision, the speed of the other. One could also measure the position of one particle, and simply calculate the position of the other.

We are admittedly making observations of one of the particles so, according to Bohr, we have affected it in such a way as to make its previously fuzzy, ghostlike properties 'real'. But this isn't so for the other particle: we have obtained its speed and position *without ever observing it*. Presumably the speed and position we deduce for that other particle is every bit as 'real' as that for the first. So what is all this talk about a particle having a 'real' speed only when an observation is made? We have also worked out its position, so why does Heisenberg say that according to quantum theory it is impossible to know both the position and speed of a particle simultaneously? Einstein concluded that the only reason quantum theory makes these statements is because it lacks something – it is incomplete.

There is a way around this: perhaps the first particle was somehow able to tell the second, 'My position has just been observed, so stop being ghostlike and get yourself a definite speed immediately!'

This smacks of desperation and, according to Einstein, it doesn't even work. He had a trump card to stop Bohr using this escape route. Suppose we wait a long time before we decide to find out where one of the particles has got to. It will then be separated from its partner by quite a large distance. Now, by Einstein's own special theory of relativity, communication between two objects cannot take place faster than the speed of light. So if we can make a measurement of position faster than the time it takes for a beam of light (or anything else) to get from one particle to the other, it seems pretty reasonable to say that the second particle *had no way of knowing* that a measurement had been carried out on its partner. The conclusion, therefore, is that both particles in fact always have concrete properties – that they are, in this sense, always 'real'.

Einstein felt confident that with this brilliant thought experiment he had finally exorcised the ghosts of 'unreal' particles with their unknowable properties from quantum theory. Particles were 'real' whether you looked at them or not. The only way around this conclusion is to say that the particles in the thought experiment are involved in some form of collusion, exchanging information faster than the speed of light – a conspiracy theory with a vengeance.

But Einstein reckoned without the ingenuity of Bohr in bringing everything into agreement with his manifesto. Using the Copenhagen interpretation, Bohr claimed, there is a way out of the EPR thought experiment.

According to Bohr, when a measurement is made on one of the two particles, it is transformed from a ghostlike entity with no definite properties into a 'real' particle. But, according to Bohr, the other, unobserved particle remains ghostlike. It therefore has no definite position. So, even though Einstein might claim to know where the other ghostlike particle is, the uncertainty principle means that its real position is completely uncertain. Hence Einstein was not justified in saying that the ghost particle lay beyond the influence of the real one – and thus he could not assume that the act of observation on the real particle has not, in fact, had some influence on its ghostly partner.

This, then, was the nub of Bohr's response: one might think that, having measured the position of *one* of the particles, you then know where *both* are using Newton's classical laws. But until you have observed them both, you have no right to say so.

Not content with merely disposing of Einstein's tricky thought experiment, Bohr went further. He introduced a view of the quantum world that has led some to see in the theory parallels with Eastern religion. Bohr maintained that the fuzziness of unobserved particles means that they are not, in fact, separate entities at all, but form a complete, *holistic* system. An observation on one can be detected by the other, even when the two seem so far apart from each other that they cannot be exchanging information.

Einstein condemned Bohr's brilliant response as an appeal to 'ghostly action at a distance'. This might not seem exactly vitriolic, but to a physicist 'action at a distance' is nothing short of witchcraft, the sort of idea only science-fiction writers call on as a *deus ex machina*. It is essentially a form of communication in which everything in the universe can influence everything else instantaneously, regardless of distance.

But after the battle over the EPR experiment, Einstein was seen by many physicists as a brilliant scientist who could no longer keep up the pace. His best work was indeed behind him and he spent his later years on a fruitless search for a 'unified field theory', a theory which would bring all of physics together in one grand synthesis. He never found it; given his stance on quantum theory, few were very surprised.

Even before Einstein's death in 1955, the Copenhagen interpretation, with all its bizarre notions, had become the established dogma of physics. Yet the echoes of Einstein's protestations can still be heard today. Is not uncertainty simply the result of our ignorance of hidden variables that would give us precise knowledge about particles, if only we knew what the variables were? Does this 'ghostly action at a distance' really exist between particles? Is there not some way round Bohr's oh-so-clever answer to Einstein's thought experiment?

OUTSTANDING MYSTERIES
Are particles really ghosts?

Most physicists would say that the great Einstein suffered a rare defeat at the hands of Niels Bohr over the implications of the famous thought experiment of the exploding molecule. Yet most physicists would also accept that a hundred victories in debate, though impressive, aren't worth as much as a single experimental verification of the phenomena at the heart of the argument.

With the Bohr-Einstein debate, however, an experimental verification appears to be asking rather a lot. Recall that Bohr's victory over Einstein was based on a distinctly unusual claim: that particles like electrons can in some sense 'tell' their partners what has happened to them. Further, Bohr was claiming that it doesn't matter how far apart the particles are – they can communicate instantaneously via the weird 'action at a distance' that Einstein found so repugnant.

But if Bohr is right, might we be able to detect the results of this interparticle bush telegraph? Clearly, any laboratory experiment set up to find them must be set up with great care, lest some of Bohr's views on the nature of reality are mixed up with some of Einstein's so that an ambiguous result is reached.

Einstein was insistent on two things. Firstly, that there can be *no* instantaneous communication between particles; the fastest any information can be exchanged is, according to his special theory of relativity, the speed of light. Secondly, he believed that the uncertainty principle of Heisenberg is just a statement of ignorance. If we knew the values of the as yet unidentified hidden variables describing the particles, we would be able to make measurements to any accuracy we liked. In other words, particles do, in fact, have well-defined qualities such as position and speed – they seem ghostlike and fuzzy only because we do not have full knowledge of the hidden variables which would enable us to calculate exact values for the position and speed. So, according to Einstein, particles are, in this sense, always 'real'.

Now it is entirely possible that, if Einstein is right and there are hidden variables which would enable us to know exactly whatever we wanted about particles, then some form of Bohr's quantum 'bush telegraph' might exist as well. According to Einstein, however, it will be limited by the speed of light.

To find out if Einstein or Bohr is right, then, our experiment must be able to tell the difference between the amount of cooperation due to hidden variables operating at the speed of light and that expected according to Bohr's view of *instantaneous* interaction with no hidden variables at all.

A bit of thought suggests that, because Einstein's communication system is limited to speeds no faster than that of light, it would not produce the same degree of cooperation between particles as Bohr's instantaneous system. This is, in fact, correct. However, it takes some very clever work to put some hard figures on the degree of cooperation expected from the two theories.

In 1964, the British physicist John Bell worked out an expression for the degree of cooperation expected according to the two rival views of quantum theory. This breakthrough opened the way to an experimental way of finding out if particles can 'talk' to each other instantaneously, as Bohr claimed.

Bell's theorem, as it has become known, can be stated thus. Let us take the three (Einsteinian) assumptions that particles are always 'real' (i.e. that ignorance of hidden variables is the reason for the uncertainty in their otherwise perfectly well-defined nature); that no communication can take place faster than the speed of light; and that the usual rules of logic still apply in the subatomic world. There is a limit to the degree of cooperation that can exist between two particles in an Einstein-Podolsky-Rosen style experiment. Bell's theorem allows one actually to calculate the degree of cooperation for a specific experimental set-up. If the experiment finds a degree of cooperation exceeding this value, then Einstein must be wrong: instantaneous communication between particles really does exist – and/or there are no hidden variables.

In the early 1970s, experiments were begun in the laboratory to measure cooperation between particles and thus to see if they could decide between Einstein and Bohr. However, each experiment had its own flaws and their results – which sometimes appeared to support Einstein – were not accepted. Then, in 1982, a team of French scientists under Alain Aspect at the University of Paris, came up with probably the most careful and sophisticated test carried out to date between the two competing views.

The experimental set-up is shown in the diagram opposite. Instead

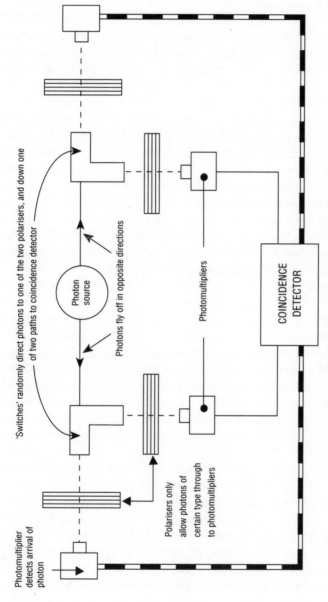

Photomultiplier
detects arrival of
photon

'Switches' randomly direct photons to one of the two polarisers, and down one
of two paths to coincidence detector

Photon
source

Photons fly off in opposite directions

Polarisers only
allow photons of
certain type through
to photomultipliers

Photomultipliers

COINCIDENCE
DETECTOR

Detector counts the number of times pairs of photons travel down
same paths – but how do they know what the other is doing?

The Aspect Experiment: proving that particles 'talk' to one another faster than the speed of light of the particle fragments of Einstein's original thought experiment, Aspect's set-up was designed to find cooperation between packets of light energy – i.e. photons – which emerge from a central source.

Every photon consists of vibrations oriented in a certain direction or 'polarisation'. If two photons emerge from the same source they must have equal but opposite polarisations. Aspect and his colleagues used this fact to help them search for signs of photons 'comparing notes' faster than the speed of light.

After each photon emerged from the central source, it was allowed to travel about six metres before being passed through an 'optical gate'. This diverted the photon in one of two directions. Either route led the photon towards a polariser – a piece of material like that found in some sunglasses which allows only photons with a certain polarisation to pass through it.

To force the photons into revealing if they really do compare notes and the extent to which they do, Aspect tricked them. The gates which dictated through which polariser the photon would finally pass were wired up so that they switched between the two alternatives randomly about 100 million times a second. Thus, while the photons were still on their way towards the polarisers, their final destination was being decided at random. A photon could not therefore 'know' in advance which of the two polarisers it would go through at the end of its flight. It would have to wait until it actually went through one or the other before it could know its final destination. And by this time it would be too late for the photon to tell its partner flying in the opposite direction what had happened to it – as Aspect had made sure that, by then, each photon would be so far apart that it would not have time to communicate with its partner even at the speed of light.

Using this arrangement, Aspect made a startling discovery. Despite their separation, pairs of photons still seemed to have a way of 'comparing notes' and telling each other which polariser they were passing through. Furthermore, the level of the 'communication' was much greater than expected according to Einstein's view of reality: five times higher in fact. It was, however, more or less exactly the amount of collaboration predicted using Bohr's views. The photons did indeed appear to be comparing notes faster than the speed of light.

So Einstein appears to have been wrong in his belief that particles cannot compare notes with each other faster than the speed of light.

In addition – or *alternatively* – Heisenberg's uncertainty principle is not just the result of particles being jiggled about in ways that could be unscrambled if we knew the value of as yet hidden variables.

By Bell's theorem, one of these two assumptions has to go. Under the Copenhagen interpretation of Bohr, it is the idea of hidden variables that is mistaken. Uncertainty in physics is intrinsic and irreducible, and is a reflection of the ghostlike nature of particles.

On the face of it, this seems to be better than the alternative: that there is no uncertainty, but that particles can communicate instantaneously with one another. That would seem to go against Einstein's special theory of relativity which forbids faster-than-light communication. This would be grave indeed, and not just because the theory has worked so well for so long. Faster-than-light signalling opens up all sorts of horrendous paradoxes, such as communication with people in the past, and thus the prospect of changing history.

But one must be very careful here. Many physicists, including Alain Aspect himself, point out that just because quantum particles can compare notes faster than the speed of light, it would be wrong to think that we could exploit this fact, and build a usable signalling device capable of bringing us up against all these paradoxes.

So was Einstein wrong in his view of quantum theory? Certainly he cannot have been completely correct: he must have been mistaken in one, at least, of his views. Particles cannot be both 'real' and incapable of faster-than-light cooperation. But deciding which of the two Einstein was wrong about is still a matter of great debate.

Exorcising the ghosts?

At the frontiers of science, researchers often have to resort to the emotional appeal of an idea to guide them. So it is with the frontiers of quantum theory. In particular, some theorists believe that the Aspect experiment should be interpreted as evidence that faster-than-light cooperation really is possible in the realm of the quantum. They find this more palatable than the prospect of never being able to overcome the uncertainty and 'ghostliness' of quantum theory today. Like Einstein, they believe that Heisenberg's uncertainty principle is overly pessimistic; the world is really a lot less uncertain and fuzzy than Heisenberg, Bohr and their followers would have us believe. In short, these theorists don't believe that particles are ghosts.

The doyen of these counter-revolutionaries is Professor David Bohm of Birkbeck College, London. Since the 1950s, he has been working towards a new approach which gets round the sort of problems theories based on quantum theory usually face.

Bohm's ideas can be traced back to some work by Prince de Broglie, who put forward a tentative hidden-variables theory in 1926. At the time, de Broglie's ideas were, in the words of John Bell, laughed out of court: 'His arguments were not refuted, they were simply trampled on.' It was yet another example of something scientists are often loath to admit exists: the Bandwagon Effect. At the reins of the quantum bandwagon were one or two superstars – like Niels Bohr – and in the back were like-minded theorists and experimentalists armed to the teeth with laboratory results to prove their case. If you get in the way of such a bandwagon, you will at best be left behind, and at worst be flattened.

David Bohm knows all about the Bandwagon Effect. As a young member of the Princeton University faculty in the late 1940s he was, ironically enough, on board the Bohr bandwagon. It was only after he started writing a textbook on quantum theory for students that he detected what he believed were gaping holes in the conventional (i.e. Copenhagen) view. Bohm started to look askance at the pronouncements of Bohr. 'His belief in the ambiguity of reality was paralleled by his statements – they were always ambiguous too,' he says. 'Most physicists said they followed Bohr, without really knowing what Bohr was doing.

'When the book was finished I began to look over the whole subject again. I was interested in the question of whether there was any way of understanding what was going on *independently* of what was being observed.' By the early 1950s, Bohm had worked up his ideas into a paper. He proposed that every particle was indeed under the influence of something described by hidden variables. The source of this 'something' Bohm termed the *quantum potential* (the 'potential' being a technical term for something that dictates how an object behaves). Bohm found that this quantum potential was uniquely defined – there was no room for fudging its mathematical form. But most important of all, quantum theory incorporating it would be just as successful as before. The only difference would be that now there was an explanation for some of the weirder philosophical points which secretly worried so many physicists.

Essentially, this quantum potential links all particles together in a huge interconnected web, enabling each to tell instantaneously what the others are up to. It is the quantum potential, says Bohm, which is responsible for letting the photons in the Aspect experiment compare notes. There need be no conflict with special relativity either: it must simply be considered valid only at a level where the source of quantum uncertainty cannot be discerned. According to Bohm, special relativity no longer works at the deeper level at which the quantum potential exerts its influence.

Bohm also claims that the quantum potential is responsible for the quantum uncertainty of particles. The potential affects particles in such a way that their well-defined classical behaviour turns into the uncertain, fuzzy quantum behaviour we observe. Thus uncertainty is *not* intrinsic, according to Bohm – it is the result of fluctuations in the quantum potential.

Bohm has even shown that his theory leads to a derivation of Heisenberg's uncertainty principle. It turns out that if one makes a measurement fast enough, or over a small enough distance, it may be possible to pin down the position or momentum of a particle more precisely than Heisenberg's relationships allow.

The quantum potential is a daring idea. But Bohm, who rubbed shoulders with some of the founders of quantum theory in the 1950s, found that the reaction of the majority of physicists to it ranged from indifference to outright hostility. 'Einstein didn't like it for two reasons; one, he could see it had some non-locality [faster than light cooperation] in it, which he never liked. Secondly, he told me that the thing was too cheap. He felt that one could get a deeper point of view [of the nature of reality]. My answer was that that may well be true, but it was better to have some point of view – that if you don't put forward some sort of view then the other side is going to take over!'

But at least Einstein had given the idea some thought. Bohm later discovered that his proposal was simply being dismissed out of hand by others. The problem seems to have stemmed from some remarks made by the influential physicist Wolfgang Pauli, who claimed to have shown that in their simplest form Bohm's ideas led to ridiculous conclusions. Bohm extended his theory and answered those criticisms but – like a scandalous but incorrect story in a newspaper – the mud Pauli had thrown seemed to stick.

Bohm heard that one professor at a major university had been quizzed about Bohm's proposal by his students, who said that they couldn't find anything wrong with it. 'But of course there's something wrong with it,' said the professor. So he looked at it and couldn't find anything wrong. Finally they brought it to Robert Oppenheimer, the brilliant and charismatic physicist who had led the development of the atomic bomb. Bohm was told that Oppenheimer's reaction was that it was 'well known' that there was something wrong with Bohm's ideas. But the students wouldn't let it go, and demanded a seminar. None of the theorists would do it, so an experimentalist was left to present a paper on Bohm's proposals. Finally Oppenheimer apparently said: 'Well, we can't find anything wrong with it, so we'll just have to ignore it.'

And that, for decades, is precisely what happened. Only relatively recently, with the death of the former Young Turks Bohr and Oppenheimer, has Bohm's proposal attracted interest again. Until his own untimely death in 1990, one of its latter-day supporters was John Bell, who heard Bohm give a lecture on the quantum potential in 1956. But, in the end, the only way to stop a bandwagon is to hit it with an unequivocal experimental test that supports a new theory. What is needed is concrete evidence that the quantum potential really exists all around us, telling particles about the world they live in and what their partners are up to. Bohm has suggested some plausible tests of his ideas. They are rather general tests and amount to looking for the existence of hidden variables by making extremely careful redeterminations of standard quantum theory quantities.

But no unequivocal backing for Bohm's theory has yet emerged. This, as Bohm has emphasised, should not be a reason for ignoring the basic ideas behind it. It took decades for suitable technology to emerge that could help decide between the views of Einstein and Bohr. What we can say, however, is that, thus far, no experiment – including that of Aspect – has succeeded in ruling out the possibility that Heisenberg was too pessimistic about the limits to knowledge. So unobserved particles may still, in this sense, be 'real' rather than ghostlike, but even if they are, we know now that they must have some decidedly supernatural abilities.

Where does the Realm of the Quantum begin?

The quantum world is truly a bizarre one, populated (at least if Bohr is to be believed) by ghostly particles talking to one another via some magical, instantaneous 'bush telegraph'. How different from our own world, with its certainty, its concrete reality and its late mail deliveries. Thankfully, we can leave all these weird ghosts and their chatter behind us when we talk about the world we inhabit.

Or can we?

Reading between the lines of the Copenhagen interpretation, one reaches a truly disturbing conclusion about just how far the Realm of the Quantum stretches.

First, recall that the Copenhagen interpretation states that until a particle is observed, it has no definite position or motion; it is everywhere and nowhere in particular. One can still use the mathematical gadget Schrödinger invented – the wave function – to describe what a particle is up to. However, to reflect adequately the vast range of possibilities open to the particle, the wave function must take the form of a vast collection of individual wave functions, one for each possibility. In the jargon, this is called a superposition of wave functions.

According to Bohr, whenever we carry out a measurement on a particle, it can no longer enjoy the run of all these possible states. It must stop being a ghost, and make up its mind what it is going to do. In other words, all these other wave functions must collapse down to just *one*, which is duly promoted to a special status that allows one to say what the particle is doing now that it is under observation.

This immediately begs a question. What is it about an observation, an act of *measurement*, that brings about this collapse of the immensely complex wave function into a single one?

To answer this deep question, which goes by the prosaic name of the Measurement Problem, a good place to start would be to try to find some aspect of whatever is doing the observing – the Geiger counter or whatever – that could exert some quantum-world influence on the particle.

Everything is made up of atoms and particles, and they can all be described by wave functions themselves. So we can think of having a quantum-theory description of *both* the observing equipment and the particle, with their wave functions overlapping and interfering with each other.

And now we see that the Realm of the Quantum, with all its wild ideas, is not restricted to the world of the atom. In trying to find an answer to the Measurement Problem, we have found ourselves having to admit that things like Geiger counters are also denizens of the Realm of the Quantum.

But where does it all end? If the Geiger counter is intimately tied to the particle it is trying to measure, then what about us? Are not humans also describable using quantum theory, and thus is not the scientist who reads the meter on the Geiger counter, which was observing the particle, not also bringing about the collapse of a wave function?

Schrödinger, the inventor of the wave function, was profoundly disturbed by the implications of everything in existence, including ourselves, being in the Realm of the Quantum. He produced a now-famous illustration of the dreadful consequences of allowing quantum ideas to slop over into the everyday world. It is another ingenious thought experiment, now known as the Paradox of Schrödinger's Cat.

Imagine an airtight steel box containing a Geiger counter, a feeble source of radioactivity, a hammer and a little bottle of deadly poison. If a single atom in the radiation source decays, the event will be picked up by the Geiger counter. This will then cause the hammer to fall on the bottle, breaking it and releasing the poison. The source of radioactivity is rather feeble. In one hour, there is only a fifty-fifty probability that one atom has decayed in it.

Now, open the box, put in a cat, and close the lid. If we wait for an hour, we can say that there is a fifty-fifty probability that an atom has decayed, thus releasing the poison and killing the cat.

But things look distinctly spookier if we use quantum theory to work out what is happening inside the box. The fifty-fifty probability of decay of the atom is then described in terms of wave functions. In particular, the state of the atom can be described by a super-position of two wave functions, describing two possibilities: the atom decayed, or it did not decay.

Similarly, the state of the cat can also be described by a wave function. This will also be a superposition of the two possible states of the cat: alive and dead.

And now we make a rather disturbing discovery. After an hour

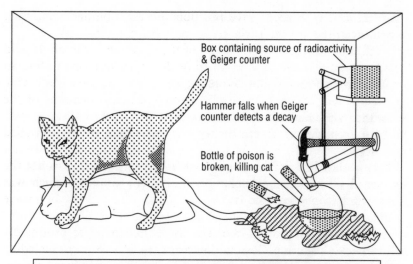

Box containing source of radioactivity & Geiger counter

Hammer falls when Geiger counter detects a decay

Bottle of poison is broken, killing cat

Quantum theory kills the cat – or does it? After 1 hour, there is a 50 per cent chance that Schrödinger's Cat is dead, but an equal chance it isn't. The Cat is thus both alive *and* dead.

has passed, there is an equal probability that it is in either state. In other words, quantum theory tells us that the steel box then contains a cat that is neither wholly alive nor dead.

But as soon as we lift up the lid to find out if this dreadful prediction is true, of course, we resolve the impasse. Our act of observation triggers a collapse of the superposition of wave functions to a single one, making the cat definitely dead or alive.

The idea of a poor cat being made 'undead' during our experiment is hard to accept. One way out would be to assume that the cat's self-observation is enough to ensure it remains forever 'real'. But what if we had used an amoeba instead? Has it enough consciousness to carry out observations to prevent itself being put into limbo until some helpful higher life form collapses its wave functions?

Again, where does one draw the line? Eugene Wigner, the Hungarian-born American Nobel laureate and leading authority on quantum theory, has championed the view that it is indeed consciousness that makes the difference. When we become conscious of something we bring about the crucial collapse of the wave function, so all the dreadful mixed states of life and death disappear.

But this has not proved a popular explanation among physicists.

Most think that Wigner's proposal puts too big a burden on to the shoulders of the less than concrete concept of consciousness.

A Quantum Theory of the Universe

Some theorists claim to have found a way of solving the Measurement Problem by taking quantum theory and applying it to the universe as a whole.

This sounds like an incredibly complex undertaking. Fortunately we don't have to take the theory very far before some very interesting things start to emerge from it.

What will a quantum theory of the universe involve? If we want to do anything interesting, we must have something to plug into Schrödinger's equation. That means we need a wave function that sums up the state of the universe.

But before we even get as far as finding one, we make a discovery. The universe includes, by definition, everything: all observers, and all possible things that might conceivably be observed. So who is going to carry out the act of measurement that brings about the all-important collapse of the universal wave function?

Two resolutions immediately suggest themselves. First, there is one observer that many people would have no problems in considering as 'outside' the universe: God. However, we very quickly run into problems with this idea. They stem from this perplexing division in quantum theory between observers and the things they observe. Let us suppose that God is responsible for ensuring that we humans are not ghostly entities. Thus He must be keeping us under permanent observation, ensuring our wave functions are collapsed to one describing concrete reality. But if He can do this, how come He doesn't also have quantum systems like electrons permanently in His divine gaze as well? Why does He leave observations of them entirely to us, who therefore have sole responsibility for converting them from their ghostlike state into reality? The 'divine hypothesis' fails because it cannot explain why God chooses to observe only part of his whole creation.

A second resolution is to take on Wigner's belief that it is *consciousness* that brings about the collapse, thus vitiating the need to go 'outside' the universe.

But the most audacious way out of the problem is that put forward by an American physicist Hugh Everett in 1957 and developed by a number of theorists since. In this, the act of observation causes our universe to split into as many copies of itself as there are possible states of what we are observing.

To take a specific example of how this many-universes picture works, let's return to Schrödinger's Cat. Here, before the act of observation forces a collapse of the cat's wave function, the cat is both alive and dead. According to Everett, the act of observation causes the universe to split into two copies, in one of which the cat is totally dead, in the other totally alive. Both also have an observer, and each observer thinks he or she has got the only picture of reality. As these two, separated in different universes, cannot talk to one another, they are never disabused of the true complexity of the matter.

So Everett's interpretation abandons the mysterious process by which the act of observation leads to the collapse of a complex wave function into a single, concrete description of reality. Every time a measurement is made, he claims, the universe splits up into many different copies. Each one of these universes contains one of the possible states into which the original wave function could be collapsed.

It comes as no surprise to learn that many physicists have problems with Everett's proposal. On the face of it, the many-universes interpretation looks like a latter-day version of the intensely contrived ideas medieval scholars conjured up to get round tricky problems. Worst of all, it seems to be completely untestable. How could one hope to detect the presence of all those other universes, which are supposedly constantly being created around us as we observe the world?

Amazingly enough, the British quantum theorist David Deutsch has suggested a possible experimental test of the many-universes theory. It requires the construction of an electronic analogue of a human brain. The components of this artificial brain will be governed principally by quantum effects and, as such, can be influenced by quantum phenomena.

According to Deutsch, it is then in principle possible to detect the co-existence of the parallel universes created by the act of observation through the interaction of their respective wave functions. If the universes exist, their wave functions would interfere with one another, creating effects that could be registered by the electronic brain.

Deutsch believes that the experiment could be carried out within the next few decades. If it did detect the presence of these other universes, the implications are literally mind-boggling.

It is hard to believe that quantum theory, with its profound challenges to our notion of 'reality', could have arisen from something as mundane as attempts to understand the nature of heat emerging from a blackened oven. Indeed, the emergence of something as revolutionary as quantum theory from such an unlikely field of research constitutes powerful arguments against those who would try to 'prioritise' science, and dump so-called 'pointless' research.

But if this brief tour of quantum theory has left you with a feeling of bewilderment, you should not doubt your own abilities to understand deep problems. You have, in fact, passed the first test of an initiate into the ways of the quantum, for, as Niels Bohr himself once said, anyone who is not shocked by quantum theory has not understood it.

It says much for the power of modern science that it can take ideas as strange as those from quantum theory and wield them into a usable theory. Doing so is far from easy – intuition, the torch which has so often lit the way to new knowledge, can be a poor guide in the Realm of the Quantum. Yet some of the brightest minds are now claiming to see the path to what has been called the holy grail of science: the theory of everything. Let us now travel down the path hacked out by these pioneers in search of the ultimate prize.

5 The Building Blocks of Creation

I N 1827, THE SCOTTISH BOTANIST Robert Brown mixed up some pollen with a little water and put the combination under his microscope. Peering down, he expected to see nothing more than the intricate shapes of the pollen grains. But he noticed that suspended in the fluid within the pollen were tiny particles performing a frenetic dance. Brown was able to rule out external causes such as evaporation for what he was seeing. So was there something alive within the pollen?

At first it appeared so – Brown found the same activity in other samples of fresh pollen. But then he made a disturbing discovery. He took a fragment of rock from the Sphinx in Egypt and ground it up to a very fine powder. Mixing the result with a little water, he put the sample under the microscope. To his astonishment, the particles of this enigmatic monument were performing the same frenetic dance.

Brown died 31 years later, still baffled by what is known to this day as 'Brownian motion'. In fact, Brown had unwittingly become a pioneer in, of all subjects, atomic physics. The particles of rock from the Sphinx we now know were being shoved about by groups of far tinier particles invisible to any optical microscope: atoms.

Atomic physics has such an aura of modernity that it is somewhat surprising to discover that the idea of an atom is a very ancient one which was spurned until the beginning of the twentieth century. Two thousand years ago, the Greek philosopher Leucippus and his student Democritus argued that in this world of change, there are certain things that remain unchanged and make up everything within it. These entities are themselves, however, supposedly indivisible, and they became known as atoms, from the Greek for 'the indivisible ones'.

Such philosophic arguments cut little ice with post-Newtonian scientists such as Brown, however. It was not until the early twentieth century and the work of, among others, the young Albert Einstein that the true significance of Brown's discovery became clear.

Since then, scientists have learnt that the word atom is a hopeless misnomer. Atoms are made up of other, even smaller particles and these, like Russian dolls, have yet smaller constituents inside them. The search for the ultimate building blocks of the creation has spawned a host of theories which try to bring some order to the bewildering variety of such particles.

Attempts to understand the forces that control the behaviour of these particles – indeed, of the entire universe – have met a similar fate. Until the 1930s, only two fundamental forces were thought to exist. There was gravity, the property of matter which somehow enables it to attract all other matter to it, no matter how far away. The other force, electromagnetism, is familiar from simple experiments such as using a plastic comb to pick up paper. Similarly, magnets were known from ancient times to be capable of picking up certain metals.

But since the 1930s, two more fundamental forces have been discovered, the so-called weak and strong nuclear forces. Scientists are now striving to bring all four of these fundamental forces, with all their wildly disparate qualities, into a single, coherent theory.

During the 1970s, an idea emerged that seemed to offer a means of bringing the overarching theories of both particles and forces together in one single, all-embracing 'Theory of Everything'. Although that grand vision is still thought to be attainable, the search for the Theory of Everything is proving to be arguably the most challenging task ever faced by scientists.

To understand just how much progress has been made in this truly heroic task, we must now delve deep into the heart of matter.

How to dissect an atom

If you want to find out what something contains, you take it apart. Scientists wanting to discover the nature of the atom have adopted this *modus operandi* with a vengeance – they smash particles into one another at speeds approaching that of light itself. 'Atom-smashing' machines like those at CERN, the European centre for nuclear research in Geneva, can reach energies similar to those reached during the birth of the universe itself.

In the early decades of this century, the pioneers in particle research, such as Lord Rutherford and his team at Manchester University, used so-called alpha particles which burst from radioactive materials as the hammers with which to break into atoms.

They, like Brown before them and many others in later years, were often astonished by what they found.

For example, in 1909, one of Rutherford's students was firing alpha particles at a target of thin gold leaf just 0.001 mm thick. His plan was to probe the innermost structure of atoms of gold by watching how they deflected the particles being shot through the foil. At the time, the consensus was that an atom was like a tiny plum pudding, with negatively charged electrons as the 'fruit' evenly spread through a positively charged 'cake mix' (see diagram on page 130).

But Rutherford's student made a very curious observation: for roughly every 9,999 alpha particles that went straight through the foil, one seemed to bounce off something and come flying straight back out. Rutherford's own reaction to this discovery is eloquent. 'It was almost as incredible as if you had fired a fifteen-inch shell at a piece of tissue paper and it came back to hit you.' Rutherford worked out what must be going on inside the gold atoms. Instead of being like some sort of amorphous plum pudding, atoms must have a central, very small core – a nucleus – that contained the bulk of the mass of the atom. Rutherford showed that this explained the deflection of the alpha particles. He was even able to come up with an estimate for the size of the nucleus – about two hundred-billionths of a millimetre.

Fascinated by this new vision of the atom, Rutherford pressed on with his experiments, firing alpha particles into a host of different targets to 'split' their constituent atoms and see what lay inside. He discovered that even the nucleus of an atom has structure to it. Experiments showed that it contained positively charged particles, now called protons. But that was not all. Although these protons were relatively heavy particles, they could not account for the total mass of the atoms to which they belonged; in fact, protons could account for only about half the mass. This led Rutherford to postulate the existence of another particle of the same mass as the proton, but which carried no charge at all – a 'neutron'. During the 1920s, he and his collaborator James Chadwick, now working together at the Cavendish laboratory in Cambridge, mounted several unsuccessful attempts to track down irrefutable evidence for the existence of the neutron. In 1932, again using alpha particles, Chadwick finally succeeded.

By the mid-1930s, physicists had arrived at the image of the atom which is taught to schoolchildren today: a central nucleus of neutrons and protons around which orbit a number of much lighter particles – electrons. Each of these electrons has a negative electric charge of exactly the same size as the positive electric charge on the protons in the nucleus. Just as opposing magnetic poles attract each other, the negative electric charges on the electrons ensured that they stayed in orbit around the positively charged protons. The force between them was electrostatic – part of the so-called electromagnetic interaction, which is one of the four fundamental forces of Nature.

Spin: the great divider

These 'subatomic' particles now became the subject of intense research, leading to further curious discoveries. It emerged that these particles were in some sense 'spinning' like tiny tops. This discovery has, as we shall see, come to play a central role in attempts to construct Theories of Everything.

Unlike the spin of an everyday object such as a football, however, this intrinsic spin can come only in certain values. The minimum is – oddly enough – half a unit of spin. All particles have spin which is either equal to this minimal value, or to some simple multiple of it. It turns out that the incredibly complex world of subatomic particles can be simply divided in two. One family of particles consists of those particles whose spin is some *odd-number* multiple of this minimum half-unit of spin (e.g. 1/2, 3/2, 5/2 etc). For historical reasons, these are known as the Fermi-Dirac particles, after two leading theorists in atomic theory. The most familiar members of this family of 'fermions', are the electron, proton and neutron. As a rule of thumb, fermions make up the matter we see around us – the tables, chairs and children.

The other family comprises those particles whose spin is an *even* multiple of the minimum (e.g. whose spin is measured in numbers such as 0, 1, 2 etc). These are the Bose-Einstein particles, and the particles which carry the fundamental forces that act on all matter in the universe belong to this family. For example, the photon (the particle which carries the electromagnetic force) and the graviton (which carries the gravitational force) have spins of 1 and 2 respectively, and are thus both 'bosons'.

So by the 1930s, scientists had a neat picture of the atom: it was like a miniature solar system, with electrons whizzing around the nucleus. They also had a tidy catalogue of the properties of the particles making up the atom, with each classified according to just three labels: mass, spin, and electric charge.

Einstein at the speed of light

In the early days of atomic physics, it was the experimentalists like Rutherford who made the spectacular discoveries. But they were soon to be eclipsed by their mathematical colleagues, wielding the equations that ruled the subatomic world: quantum theory.

As we learnt in the last chapter, by as early as 1913 quantum theory had given birth to an explanation of why electrons do not simply spiral into the nucleus, destroying atoms and giving the likes of Rutherford nothing to work on.

By the mid-1920s, Erwin Schrödinger had brought out his famous wave equation, which showed that quantum theory could account for many of the discoveries of Rutherford and his contemporaries. But impressive as this was, it was clear that Schrödinger's equation was far from quantum theory's last word on the nature of the atom. This was because, in working out the form of his equation, Schrödinger had left out one of the cornerstones of modern physics, introduced by Einstein twenty years earlier: the special theory of relativity.

As with quantum theory, the origins of special relativity are almost humdrum when compared to the theory to which they gave birth. The nineteenth-century saw scientists such as Michael Faraday in England, Hans Oersted in Denmark and André Marie Ampère in France carrying out experiments that opened the way to the technology that underpins modern society. For example, Ampère (who reputedly mastered all of mathematics then known by the time he was twelve) showed that electric currents had magnetic effects, while bookbinder-turned-physicist Faraday, working at the Royal Institution in London, showed that moving a magnet in and out of a coil of wire induced a flow of electricity in the wire. This latter work led to today's huge electricity power stations, where the rotation of steam-propelled turbines creates current for the national grid.

In the 1860s, the Scottish physicist James Clerk Maxwell performed something of a mathematical miracle by summing up all these myriad discoveries in a single set of equations. He showed that electricity and magnetism are in fact different aspects of one single phenomenon: 'electromagnetism'. But his equations also made a prediction: there should be waves of electromagnetic energy which travel at a fixed speed, apparently equal to the speed of light. This astonishing prediction – verified later in the nineteenth-century by Heinrich Hertz in Germany – was the first glimmering of the existence of radio waves, microwaves and x-rays so familiar to us today.

But there was a major problem with this brilliant work of synthesis. The answer you get from Maxwell's laws depends on whether or not the phenomena you are trying to understand are moving relative to you. Such a vital qualification is hardly what one expects from a theory which is purporting to have universal validity.

Things came to a head in the 1880s, when two American physicists named Michelson and Morley carried out an experiment whose results seemed to contradict common sense. They discovered that, contrary to expectations, the speed of a beam of light does *not* depend on how the source of the light moves. This seems at least a little odd. Surely when one measures the relative speed of something which is hurtling past, and then again when one is keeping pace with it (i.e. the relative speed is zero), the answers should be different?

Einstein learned of the Michelson–Morley experiment while still a student. With an intuition that defies explanation, he realised that he should take the result at face value. That is, he decided that a new law of Nature had been discovered: the speed of light really is the same for all observers, whether stationary or moving steadily.

Incredibly, by following through the relatively simple mathematical consequences of this idea, Einstein uncovered a new concept of space and time. The centuries-old view that there are absolute standards of length and time – ultimate rulers and clocks against which all else can be measured – had to go.

Put a clock aboard a fast-moving rocket, and send it off into space at a speed close to the speed of light. Einstein's equations revealed that compared with a clock kept on Earth, the moving clock will run slow. Neither clock is right or wrong; time kept by clocks is simply relative, not absolute – hence the term 'relativity'. Rockets travelling at high speed also shrink slightly relative to their Earthbound counterparts.

Such curious effects do not show up in everyday situations – the speeds involved are nowhere near the 300,000 kilometres per second of a beam of light. In such circumstances, Einstein's equations give results that are virtually identical to those expected from common sense. But they have been measured directly. For example, in the late 1960s, subatomic particles called pions were accelerated to speeds close to that of light and were found to last longer before decaying than their stationary counterparts.

Einstein's special theory of relativity has two implications of great importance for all modern theories of physics. The first is that it leads to the famous relationship tying mass to energy: $E = mc^2$. E is referred to as the total energy locked up in the mass, m; it represents the maximum amount of energy one can extract by any process from a given quantity of matter. What makes this relationship so important is the appearance of c^2, the speed of light squared. The speed of light is such a huge number, and its square so much larger still, that the total amount of energy locked up in a kilogram of material would be enough to keep a 100-watt bulb burning for 29 million years. In fact, conventional processes for extracting energy from matter never get anywhere near this maximum value; even nuclear reactions within the sun release just 0.7 per cent of the maximum energy available from the raw fuel.

The other major value of special relativity is that it has the power to give any theory of physics enormous applicability. Specifically, by building special relativity into a theory, one can guarantee that the theory will give the right answers, whether they are applied to situations that are stationary or moving steadily. And this, of course, was the key problem with Maxwell's otherwise wonderful laws of electromagnetism. Once Einstein's equations from special relativity were built in, Maxwell's laws gave the right results for objects moving at any velocity – up to and including the speed of light.

QED: the most successful theory ever devised

Plug special relativity into quantum theory and you have a theory which will have similarly broad applicability. But this is easier said than done. Schrödinger's equation is complex enough without having to put in relativistic considerations as well. Indeed, physicists still fight shy of putting special relativity into their equations if they can justify keeping it out – by, for example, arguing that the speeds involved in a particular problem never approach that of light.

But it soon became clear that, as far as quantum theory was concerned, physicists could not keep their heads in the sand forever. Special relativity simply has to be included in any attempt to understand the way in which matter interacts with electromagnetic radiation such as radio waves – which, of course, travel at the speed of light. In 1928, a diffident, somewhat enigmatic young British physicist named Paul Dirac wrote down the complex mathematical equation that brought together the theory of electromagnetic phenomena, special relativity and quantum theory for the first time.

Even among theoretical physicists – a breed renowned for their otherworldliness – Dirac was a singular personality. His diffidence apparently stemmed from his somewhat stern Swiss father's insistence that the boy Dirac speak to him only in French. Dirac thus became famously laconic. Tutorial students at St John's College, Cambridge, where he did his most famous work, jokingly referred to a new unit of data transfer: a Dirac, or one word per minute.

Dirac's 1928 paper did, however, make him the talk of the physics world. Because it deals with the motion (dynamics) of particles (quantum entities) in the presence of an electromagnetic field, it is called quantum electrodynamics – QED for short. It is the most successful scientific theory ever devised.

Although it took some time to turn Dirac's original work into the fine version of QED being used today, the theory did pretty well from the outset and won its originator a share (with Schrödinger) of the 1933 Nobel Prize in physics, at the very early age of 31.

One of QED's first successes was in providing an *explanation* of the spin of a particle. Until the advent of QED, spin had been regarded as a property like mass and electric charge which particles just happened to have. Scientists tend not to ask *why* an electron has a mass about 2,000 times less than a proton – it just does. But Dirac showed that the property of spin could be *derived*.

The discovery of this connection between the abstract world of mathematics and the world we see around us is an example of what scientists mean when they talk about unravelling the mind of God. The aim is always to find theories that do more than simply describe and label the universe. Dirac's theory does not simply describe what happens if an electron has spin, it actually *demands* that an electron has spin. The physicist has no choice in the matter.

Dirac's theory held further unexpected surprises in its mathematics. Like the quadratic equations still taught at school, Dirac's equations could be interpreted as having not just one but two solutions. One solution described the behaviour of an electron but the other described the behaviour of something new: a particle with exactly the same mass, but opposite electric charge.

Dirac's terseness belied the courage – almost arrogance – that accompanies so many of the greatest scientists. A less daring theorist than Dirac might have simply thrown out the second solution to the equations as an unwanted figment of the mathematics. But in 1931, Dirac announced that his equations were telling physicists something important – that for every particle, there exists an 'antiparticle'. For the electron, the antiparticle is called the positron – its name reflecting the fact that it has a positive charge, exactly opposite to that on the electron.

Dirac's boldness was quickly rewarded. The following year, a 27-year-old scientist named Carl Anderson at the California Institute of Technology was studying the tracks of particles that had zipped through the Earth's atmosphere from deep space. While some looked like the tracks expected from passing electrons, others seemed strange – they curved in exactly the opposite way to the tracks of the electrons. This showed that the particles were identical to electrons in everything but electric charge. Anderson had found Dirac's positrons.

Why don't all atoms explode?

The brilliant success of Dirac's theory convinced physicists that despite its appalling complexities, relativistic quantum theory must play a key part in a comprehensive account of reality. But while the theoreticians were rejoicing in their latest successes, some embarrassingly basic questions remained unanswered. Although quantum theory had provided an explanation of how electrons could circle the nucleus of an atom without crashing inwards (see page 125), there remained the mystery of why the nucleus existed at all.

Experiments had shown that the nucleus comprised a collection of neutrons and protons. But protons all carry positive charges – and similar charges repel one another. So there must be another, rather strong, force in the nucleus which stops the mutual repulsion of the protons from blowing the nucleus to bits.

What could this 'strong nuclear force' be? One candidate could immediately be ruled out: gravity. Given the tiny mass of the proton, gravity is far too weak to overcome the mutual repulsion caused by the electromagnetic force. Experimentalists gave the theorists an important clue to the nature of the strong nuclear force: it was short-range, effectively losing all its strength just a few million-billionths of a metre from the nucleus. This was something new: the familiar electromagnetic and gravitational forces both enable particles to affect one another, albeit very slightly, throughout the universe.

In 1935, the Japanese theoretician Hideki Yukawa showed that this peculiar feature of the strong nuclear force leads not just to an insight into a new fundamental force of nature, but to a new way of thinking about *all* the fundamental forces. As such, Yukawa's work ultimately paved the way for today's attempts to produce a theory of everything.

How particles carry forces

Most people have a vague mental picture of gravity or the electromagnetic force as some sort of invisible elastic band. Yukawa's theory leads to an entirely different – quantum mechanical – view of forces. According to this, all the forces are ultimately the result of special force-carrying particles flitting from one place to another.

This isn't as weird an idea as it might seem at first. To give some mental picture of how it works, think of two children, Adam and Bill, skateboarding together down a road (see diagram opposite). Adam has a ball which he throws quite hard at Bill. In throwing the ball over towards Bill, Adam is himself thrown backward slightly. This is just Newton's rule that to every action there is an equal and opposite reaction. Bill, in catching the fast-moving ball, is pushed away from Adam as the momentum of the ball is passed to him. So, to someone watching nearby, Adam and Bill have succeeded in repelling one another by exchanging the ball – just as if they were under the influence of some repulsive force.

By changing – and straining – the analogy slightly, it's also possible to mimic an attractive force. This time, Adam throws a boomerang in such a way that it loops over his head and travels to Bill, with the

How to picture forces due to particle exchange

Adam

Bill

'Repulsive force'

By throwing the ball, Adam is pushed further from Bill by reaction to the action of the throw. Bill also moves further from Adam, as by catching the ball, he absorbs its momentum. The analogy also shows that the heavier the ball, the shorter the range over which the 'force' can operate – (it's harder to throw!)

laws of motion this time forcing Adam and Bill together. The analogy even predicts that the heavier the object Adam has to throw at Bill, the shorter range the force – Adam simply can't throw a cannonball as far as a rubber ball.

It turns out that Yukawa's theory accounts for the strong nuclear force which stops every nucleus exploding by means of the exchange of force-carrying particles called mesons. Yukawa used the experimental data on the short range of the force to estimate the mass of these carriers of the strong interaction. He reckoned that mesons should have a mass about 270 times greater than that of the electron.

Experimentalists began to search for mesons and very soon Anderson, the discoverer of the positron, found a particle that seemed to fit the bill. It turned out that Anderson had in fact found the muon, a particle with about the right mass but otherwise unlike Yukawa's predicted carrier of the force. Finally, in 1947, during experiments on cosmic rays pouring through the atmosphere from space, the British physicist Cecil Powell found Yukawa's particle. It is now called the pion ('pie-on'), a contraction of its original name of pi-meson.

The four forces take their places

With the strong nuclear force now firmly established as a fundamental force of nature, many puzzles about the atom – including its very existence – fell into place. But there still remained an annoying gap in this new understanding – the curious case of the vanishing neutrons.

The centres – nuclei – of certain atoms, such as a type of potassium atom found in the Earth, are unstable. In their search for stability, these nuclei perform a clever conjuring trick: they turn neutrons into protons. This requires some ingenuity on the part of the nuclei. Neutrons have no electric charge on them, but protons do – and electric charge at the beginning of the decay must always be the same at the end. Sure enough, when the proton pops into existence in the place of the neutron in these nuclei, it turns up with an electron in tow, whose negative charge neatly cancels out that on the proton.

This curious process is known as beta decay, after the 'beta particle' label that electrons were originally given by Rutherford. And the more physicists learned about beta decay, the stranger it seemed. Most disturbing of all was the fact that the speed of the electrons whizzing out of the transformed nucleus could not account for the whole amount of energy involved in the process. Physicists are always loath to give up one of their most cherished principles – in this case, the principle that that mass or energy cannot be created out of nothing. So in 1931 the Austrian physicist Wolfgang Pauli proposed that the law of energy conservation is still true – it's just that with beta decay, some of the energy escapes from the nucleus in the form of a new, *unseen* particle.

To keep everything else in order, this particle had to have no charge and hardly any interaction with ordinary matter – hence its elusiveness. The Italian theorist Enrico Fermi therefore dubbed the new particle the neutrino – 'the little chargeless one'. It finally turned up in experiments carried out in 1956, more than twenty years after Pauli's brave prediction.

Physicists also found that neutrons could perform their quick-change trick outside the nucleus. Left to themselves neutrons fall to bits in about 1,000 seconds, leaving behind protons, electrons and antineutrinos. Theoretical work proved that this was symptomatic of the fourth and final fundamental force to emerge, one far more feeble than either the strong or electromagnetic forces. This is the so-called weak nuclear interaction.

With the discovery of the weak interaction, physicists knew the nature of the forces responsible for everything from the lighting of a match to the birth of the universe. Strongest of all is the strong force which binds the nucleus together. Its strength accounts for the huge amount of energy that can be liberated in nuclear fusion reactions in the sun and stars – and in hydrogen bombs. Next in strength is the familiar electromagnetic interaction, which enables magnets to pick up pins and charged combs to pick up bits of paper. Then comes the weak interaction, which accounts for spontaneous radioactivity and the break-up of neutrons. Finally, and by far the weakest, comes gravity – the force which is holding you to the ground.

The infinity disease rears its head

One would think that having uncovered the existence of the four basic forces that control the universe, physicists would simply tidy up a few loose ends here and there and then take themselves off for a well-earned mass retirement. But Nature has never given scientists an easy time, and so it proved with the understanding of the funda-mental forces and particles.

The first clouds appeared on the theoreticians' horizon following Dirac's mathematical *tour de force* of combining Schrödinger's equa-tions and special relativity to produce quantum electrodynamics. QED is at root a theory of how electrons and radiation such as light interact with one another. Expressed like this, it sounds rather abstruse. But the room you are sitting in is filled with electrons in materials interacting with light – if it weren't you couldn't be reading about QED. The distinguished English theoretician Freeman Dyson puts the theory's significance into clear perspective. He calls QED the 'theory of the middle ground' – a mathematical account of the basic physics behind everything from the atomic nucleus up to an entire planet.

Dirac's original conception of QED had some impressive early successes – such as explaining why electrons have spin – but by the late 1940s it was beginning to run into trouble.

The problems lay in the way Dirac's theory dealt with electrons. His version of QED predicted that the electron behaved in some respects like a tiny magnet. In particular, his equations said that a property that can loosely be called the strength of this magnet was exactly one 'strength unit'. But around 1948, experiments discovered that the real strength was about 1.00118 units – *almost* exactly, but not quite, the same. Aficionados of Dirac's theory could not blithely write this off as just sloppy work by their experimentalist colleagues. The experiments were accurate to about three parts in 100,000 but the discrepancy between theory and experiment was 118 parts in 100,000 – so there was little doubt that something was wrong with the theory.

Ambitious theoreticians set about trying to make more accurate predictions of the magnetic strength of the electron, this time taking into account more subtle effects of QED that had been hitherto ignored. The trouble was that the results of these refinements turned out to be mathematical gibberish. Instead of getting something like 1.00118, the calculations gave a value for the strength that was *infinitely* big – rather larger than the measured value.

This astonishingly wrong answer signalled an outbreak of 'infinity disease', a mathematical sickness whose symptoms are the appearance of infinities in the results of otherwise straightforward calculations. Infinity disease has come to be dreaded by theoreticians, and has plagued attempts to formulate grand theories of the universe for decades; we shall run into it again.

Fortunately for QED, an antidote for this outbreak was soon found. In 1943, working in war-torn Japan in utter isolation from the mainstream of physics, Sin-Itiro Tomonaga appears to have found the answer first, although it was not until after the war that his work reached the rest of the scientific community. By that time, two Americans had also independently found ways of avoiding the infinity disease. Julian Schwinger, a child prodigy who was a professor at Harvard by the age of 27, produced an intensely mathematical way around the problems.

But at Cornell University, a wild and wonderful young theorist named Richard Feynman had developed a radically new approach to the QED problem, based on what have become known as Feynman diagrams. Like his approach to the QED problem, Feynman was always springing surprises. As a child, he developed his own mathematical notation, and made a point of never taking anything in a textbook as read. Feynman emerged as one of the greatest physicists of the twentieth century, whose personal life was as quirky as his approach to physics – his outrageous autobiography *Surely You're Joking Mr Feynman!* contains advice on how to pick up women and pick safes.

To start with, no one could see how Feynman's weird diagrams could be remotely related to Tomonaga and Schwinger's mathematical hieroglyphics. Eventually all three techniques were shown to be essentially the same, and all three won a share of the 1965 Nobel prize for physics. In the long run, however, it was Feynman's approach that everyone adopted. Essentially, the cure for infinity disease turned out to be a mathematical sleight of hand called 'renormalisation'. This impressive-sounding name belies the fact that many physicists view it with a considerable amount of scepticism. One does not have to be a Nobel prizewinner to see why. Essentially, to get rid of an infinity, one waters it down to a more reasonable value, according to some rules. The odd thing is that when this is done in QED, the number left over turns out to be just the value for the magnetic strength of the electron found in the experiments.

Physicists thus found themselves behaving like witch doctors. Their theories have a 'disease', and they know that a treatment called renormalisation works – but they don't really know why. Still, the 'patient' gets to live to fight another day, so why ask questions?

But is renormalisation just a fiddle? Dirac himself thought the whole thing stank. He put his views trenchantly in 1978: 'Most physicists say that QED is a good theory. I must say that I am very dissatisfied with the situation because this so-called "good theory" does involve neglecting infinities which appear in its equations, neglecting them in an arbitrary way. This is just not sensible mathematics. Sensible mathematics involves neglecting a quantity when it is small . . . not just because it is infinite and you do not want it.'

To this day, ways of avoiding renormalisation are still being actively pursued. But most theorists are happy to close their eyes to the obvious problems and adopt the pragmatic approach: if it works, do it.

Renormalisation certainly did QED a power of good. Physicists found they could do more than just get sensible answers out of the theory – the results were just *astonishingly* accurate. The situation was summed up by Richard Feynman in 1985, in his own inimitable style: 'At the present time I can proudly say that there is no significant difference between experiment and theory! Just to give you an idea of how the theory has been put through the wringer, I'll give you some recent numbers: experiments have Dirac's number at 1.00115965221 (with an uncertainty of about 4 in the last digit); theory put it at 1.00115965246 (with an uncertainty of about five times as much). If you were to measure the distance from Los Angeles to New York to this accuracy, it would be exact to the thickness of a human hair. These numbers are meant to intimidate you into believing that the theory is probably not too far off!'

With QED back on the road and running very well, the quantum theorists were in fine fettle during the early 1950s. Meanwhile, their experimentalist colleagues were beginning to get into something of a panic.

An embarrassment of riches

Back in the good old days of Lord Rutherford, there had been just a few particles for the experimentalists to play with – the electron, proton and neutron. But then the floodgates started to open. Yukawa started it off, with his prediction of the meson. The search for this turned up another one no one was looking for, called the muon. Then the experimentalists started creating their own problems, with the invention of machines that could reach energies high enough to smash known particles together, and see what the results were. The result was yet more new particles.

These particle accelerators, such as the ones at CERN in Geneva, were almost too successful. By the early 1960s, the dream of a simple universe with a few particles able to account for everything looked to be just that – a dream. Literally dozens of new particles were turning up, so many that physicists seemed to run out of sensible names and had to give them strange monikers like Sigma-minus and Xi-star. Peter Kalmus, a distinguished British experimental particle physicist, recalls that at the time some wag suggested that instead of getting prizes, the discoverers of these new particles should pay a fine.

Some of the new particles had strange properties to match their strange names. For example, some, like the Sigma, burst out of strong interactions with nuclear material and then seemed almost pathetically to dissolve slowly away, lasting far longer than theorists expected. The American theoretician Murray Gell-Mann and some others decided that such strange behaviour betokened a whole new property of matter, which should take its place with the familiar ones of mass, charge and spin. This new property was dubbed, appropriately enough, strangeness.

This new property brought a little more order to the increasingly odd collection of particles turning up in the accelerators. But there was no getting around the fact that the sheer numbers of particles were getting out of hand. It was time to put the house in order. Physicists began by gathering all the particles together and putting them into various families. The most fundamental division of particles was made on the basis of spin. All particles either have whole-number amounts of spin – as do the so-called bosons, or fractional amounts – the so-called fermions.

Then each of these big families was divided up. Fermions were subdivided up into two groups, according to their mass. Relatively heavy fermions such as protons and neutrons, and the more esoteric particles such as the Sigma-minus were collectively dubbed *baryons*, after the Greek word for 'heavy'. Relatively light fermions, like the electron and neutrino, were called *leptons*, after the Greek for 'slender'.

But this was not the only way of classifying the particles. It was possible to divide them up according to the fundamental forces to which they responded. For example, particles such as neutrons and protons that feel the effect of the strong nuclear force were dubbed *hadrons*.

It was clear that just classifying these particles was not going to be simple. Then, in 1960, a breakthrough occurred. Murray Gell-Mann and the Israeli physicist Yuval Ne'eman independently noticed something very intriguing about all these hadrons. By plotting the strangeness of various particles on a graph against their electric charge, neat geometrical patterns emerged (see diagram on page 172). When a theorist comes across something like that, it is hard to believe that one has not been let in on some deep secret – that one has, in some way, been granted a rare glimpse of the mind of God.

Whether Gell-Mann and Ne'eman had been so privileged could quickly be put to the test. The mesons, including Yukawa's pions, formed a hexagon when their strangeness and charge were mapped out on a graph; so did some of the baryons. Yet, intriguingly, the patterns weren't quite complete. The hexagon pattern of the baryons consisted of eight particles – six at each corner, plus two in the centre. The hexagon pattern of the mesons, on the other hand, had only one particle at the centre – the pion invented by Yukawa to transmit the strong nuclear force. Were these geometrical patterns hinting at an as yet undiscovered particle, whose properties would put it next to this pion at the centre of the hexagon?

In fact, there was more: one of the hexagon patterns for the baryons seemed to be part of a bigger pattern – a triangle. But only two of the three points of the triangle had known particles sitting at them.

Surely Nature could not have resisted completing these two patterns? Gell-Mann and Ne'eman decided not, and worked out a theory for the eight-particle patterns, which became known as the Eightfold Way. The conclusion of this work was that the gaps in the patterns really did represent undiscovered particles. Sure enough, in 1961, a particle with identical charge and strangeness to the pion, thus putting it next to that particle in the meson hexagon, turned up; it is now called the eta meson.

Would the triangle pattern also be filled? Flushed with the success of the hexagon, Gell-Mann (with Ne'eman not far behind) predicted the existence of an undiscovered baryon which would sit at the third point of the triangle pattern. Gell-Mann used the pattern to predict that the particle, which he dubbed Omega-minus, would have the same negative charge as the electron, a strangeness of minus three units, and a mass 3,288 times that of the electron. In 1963, the particle accelerators at Brookhaven laboratory in New York announced the discovery of a particle with negative charge, a strangeness of minus three units, and a mass of 3,286 times that of the electron – a stunning confirmation.

The prediction and discovery of the Omega-minus is a definitive example of pure science at work: clean, clear-cut theory and prediction, and the sort of outcome most scientists only dream about. But more than that, it showed that physicists were right to put faith in so simple and beautiful an idea as symmetry. The Eightfold Way kindled a light that has helped theorists see their way through appalling complexity ever since.

The atomic Russian doll is opened

Gell-Mann didn't stop with the Eightfold Way – he wanted to know why it works. Why do the beautifully simple patterns exist? By 1964, he and another American physicist George Zweig had realised that the patterns could be explained if baryons and mesons weren't 'elementary' particles at all, but were the result of *combinations of other particles 'inside' them*. Gell-Mann conjured up the name 'quarks' for these sub-subatomic particles. (Curiously, he later noticed that James Joyce had already invented the word quark, in *Finnegans Wake*: 'Three quarks for Muster Mark'.)

According to this new picture of matter, atoms – 'the indivisible ones' – were really miniature Russian dolls, with subatomic particles inside them, and quarks inside some of the subatomic particles. But the quark proposal proved enormously powerful in bringing law and order to the disparate families of particles. Just three types of quark could account for the properties of every one of the 30 or so hadrons discovered by the early 1960s.

The quarks were given whimsical names: Up, Down and Strange. Like electrons, each quark also had an antiparticle partner: anti-Up, anti-Down and anti-Strange. To make your desired particle, you had to form the right combinations of these three quarks and antiquarks (see diagram overleaf). The general rule is: to build a meson (such as a pion), combine a quark and an antiquark. To build a baryon like a proton, combine three quarks.

Following these rules, one ends up with results in neat accord with what emerges from particle accelerators. But the theory behind the quarks shows that they themselves are rather curious creatures. They carry less than one whole unit of electric charge. Gell-Mann knew, however, that no experiment had ever seen a particle with less than one full unit of charge (and this holds true to this day). In an echo of the bitter turn-of-the-century debate about the reality of then-undetected atoms, some physicists started to wonder if quarks were anything more than mathematical devices.

The discoveries of the particle accelerators strongly suggested otherwise. By the late 1960s, the number of known particles had swelled to over one hundred. Yet, incredibly, every one could be accounted for using the quark hypothesis.

The only thing lacking was some decent evidence that quarks actually existed. Worryingly, all experiments to detect 'free' quarks floating outside the confines of the hadrons failed. The alternative approach was to go looking for them where they were supposed to be – buried deep in the heart of subatomic matter.

The particle-building game and The Eightfold Way

To build a *baryon*, such as a proton, take three quarks:

D + 2U = proton

'DOWN'

'UP'   'UP'

To build a *meson*, such as a negatively charged pion, take a quark and an antiquark:

D + Ū = pion

'DOWN'   'ANTI – UP'

By classifying particles according to their properties
(e.g. electric charge and 'strangeness'), curious patterns emerge:

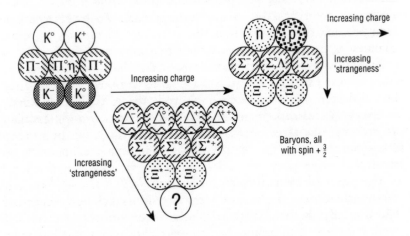

Increasing charge

Increasing 'strangeness'

Increasing charge

Increasing 'strangeness'

Baryons, all with spin + $\frac{3}{2}$

Symmetry of triangle suggests the particle at the bottom should have
negative charge, and high strangeness. In 1964 the Ω^- particle was found
in accelerator experiments – with just these properties.

The hunting of the quark

Years before quarks were even dreamt of, Robert Hofstadter and his colleagues at Stanford University in California had been firing electrons at protons and neutrons. They found that protons and neutrons aren't pointlike objects but fuzzy blobs about 10^{-15} metres across – a discovery that won Hofstadter a share in the 1961 Nobel prize for physics.

Hofstadter decided to probe further and map out the shape of the blobs in greater detail. To do this, a much more powerful electron 'gun' was needed, and just such a machine was completed in the early 1960s at Stanford: the two-mile-long particle accelerator based at the Stanford Linear Accelerator Center.

At first SLAC concentrated on improving on Hofstadter's pioneering work. But by 1968, the machine was switched to a much more exciting task: the hunt for the quark. At the end of the SLAC machine's 'muzzle', scientists from Stanford and the Massachusetts Institute of Technology set up a tank of liquid hydrogen. They then blasted electrons into the tank so hard that any protons hit by the electrons were smashed to bits. According to the theorists, any quarks hiding inside the protons should then reveal themselves by subtle changes in the flight of the electrons after impact.

And so it proved. Careful analysis of the trajectories of the electrons revealed that they were being deflected by tough little 'nuggets' buried deep in the protons. Quarks really did exist. Then at CERN in Geneva, experimenters put the icing on the cake by firing neutrinos at protons and showing that the particles inside the proton had fractional charges of exactly the values predicted by the quark hypothesis.

The proof of the existence of the quark was a triumph for both the imagination of the theorists and the skills of the experimenters. Most of the principal characters involved won Nobel prizes. But why were the stars of the show, the quarks, so reluctant to show themselves? The particle-accelerator data provided a possible explanation: according to the SLAC results, protons contained more than just quarks. They seemed to contain other, electrically neutral particles. Theoreticians soon started talking of these other particles acting as some sort of 'glue' within the proton, preventing the quarks from escaping. They dubbed them gluons, a name which has stuck.

During the 1970s, a full-blown theory for the action of these gluons was built up. It is graced by the delightful name 'quantum chromodynamics', or QCD. The 'chromo' part (from the Greek for 'colour') comes from the central idea of QCD: that these gluons prevent the quarks from escaping by gripping them by the so-called 'colour force'. To do this, the colour force has to have a decidedly odd property: it actually gets *stronger* the further the quarks are from one another – which is decidedly anti-intuitive.

But QCD does more than describe the action of quarks and gluons. Because quarks are constituents of the protons and neutrons, QCD is central to any understanding of the atomic nucleus. In particular, the strong interaction which stops every nucleus disintegrating is now seen as a complex manifestation of the colour force. Having served its purpose for many years, Yukawa's revolutionary pion-based explanation for the strong interaction has now been superseded.

The charm of the atom

By the early 1970s, the chaos threatening particle physicists in the late 1950s seemed like a bad dream. There were just two basic types of particle to worry about: the lightweight leptons, such as the electron and the neutrino, and the quarks. Almost unbelievably, all the other particles in the universe, from the familiar proton to the strange Sigma, could be made up of the three known quarks: Up, Down and Strange.

Yet many physicists realised that this state of happy simplicity could not last. You did not have to be very good at mathematics to work out that the discovery of just one more hadron by the accelerators would bring the elegant edifice crashing down. The reason was simple: all the permutations of quarks and antiquarks had been used up. There just wasn't room for any more hadrons. And in 1974, disaster duly struck: within days of one another, two teams in America discovered another hadron. The particle became known as J/psi.

The obvious way out was to conjure up another quark to give a new, even bigger, set of permutations of quarks and antiquarks. In fact, the idea of a fourth quark had been raised ten years earlier, by James Bjorken and Sheldon Glashow in America. It was not just a flight of fancy, either, but was based on the guiding principle of particle physicists: symmetry. They pointed out that there are *four* leptons: the electron, the muon and two types of neutrino, the electron neutrino (the relatively familiar one) and the so-called muon neutrino. But just *three* types of quark seemed to be able to account for all the particles then known. Could there be a fourth quark, giving one to match every lepton? Bjorken and Glashow thought so, and even gave it a name: the 'charmed quark'.

In 1970, Glashow and his colleagues published what they considered to be impressive, but still indirect, evidence for the existence of the charmed quark. What was needed was direct evidence from a particle accelerator. Yet four years later, the experimentalists still had no evidence for charm. By this time, Glashow was beginning to get a little annoyed with them. At a meeting of experimentalists in Boston that year, Glashow said that he would eat his hat if the charmed quark had not been found by the next meeting, scheduled for 1976. Fortunately for Glashow, teams led by Sam Ting at Brookhaven and Burton Richter at Stanford independently found the J/psi particle in November 1974. Less fortunately, however, J/psi turned out to be made up of two quarks, Charm and anti-Charm. As such, psi doesn't actually have any 'charm' itself; that property is cancelled out. Thus there was initially some doubt about whetherJ/psi really did contain 'charm'. But by the summer of 1976, another particle – the so-called D-meson, a combination of an Up quark and the new Charm quark, had been discovered. The property of 'charm' was now clear for all to see. By the 1976 meeting of experimentalists, all doubt had disappeared. The organisers celebrated by giving everyone a sugar hat to eat.

Just one year later, the particle accelerator at Fermilab in Chicago produced another particle, dubbed Upsilon, that not even the four quark varieties could cope with. The theoreticians this time had no hesitation in hitting back with another type of quark, called Bottom.

One big happy family

By the end of the 1980s, a surprisingly clear picture of the subatomic world had emerged from the mess of thirty years earlier. The current view is that the fundamental building blocks of matter in the universe are the leptons such as the electron, and quarks such as Charm. The idea that for every quark there is a lepton has been largely confirmed, and where certain particles are missing, physicists are sure the gaps are about to be filled. There are now thought to be, in total, six quark varieties, of which five have so far been discovered. These are Up, Down, Charm, Strange and Bottom. The sixth, called Top, is now on every experimentalist's 'most wanted' list of particles.

Five leptons are also now known to exist, with four of them forming tidy pairings of properties: the electron and electron neutrino and the muon and muon neutrino. The tau particle is the fifth lepton, and excellent – though still indirect – evidence exists for its partner, the tau neutrino.

We can now give a fairly confident answer to a very fundamental question: 'How many particles are needed to make up all the matter in the universe?' The answer – at least for the moment – is twelve: six quarks and six leptons. It is, of course, entirely likely that the huge particle accelerators now operating or under construction will throw up something to wreck this neat picture. At CERN in Geneva, plans are being drawn up for the Large Hadron Collider, which will smash protons into protons at energies that existed in the first million-millionth (10^{-12}) of a second after the birth of the universe. Under farmland near Waxahachie, Texas, American scientists have plans to build by the end of this century a truly awesome accelerator 86 km in circumference – the Superconducting Supercollider (SSC) – that will reach even higher energies.

These machines will cast new light on the nature of matter, perhaps finding the elusive Top quark. But there is more to the universe than just the particles of matter. There are also the particles that carry the forces. QED tells us that the electromagnetic force is the result of packets of electromagnetic energy – photons – flitting from one electrically charged particle to another. Quantum chromodynamics ascribes the strong nuclear force holding the nucleus together to the effects of the colour force between quarks and the gluons that flit between them.

But there are two other fundamental forces we have yet to deal with: the weak interaction, responsible for certain kinds of radio-active disintegration, and gravity. Can we understand these, too?

The strange case of the fogged plates

In March 1896, a French physicist opened a cupboard inside his laboratory in Paris and got a surprise. Henri Becquerel, youngest in a long line of scientists in his family, had just been made a full professor at the famous École Polytechnique, and was in the middle of some fairly mundane research into the phenomenon of phosphorescence – the ability of some compounds to glow in the dark.

Becquerel was particularly interested in the phosphorescence of uranium compounds, which seemed to be particularly powerful sources of the effect. The usual way of triggering phosphorescence was to expose the material under study to high-frequency ultraviolet light. Uranium crystals exposed to sunlight proved capable of fogging photographic plates even through a piece of metal.

But in early March, Becquerel developed plates that had been locked up in the cupboard with the uranium for days. Kept in the dark for so long, the crystals should have lost their abilities to fog the plates. But the plates were heavily fogged. The uranium seemed to be emitting rays of something more than just phosphorescence. Becquerel had, in fact, just discovered a new property of matter: radioactivity.

To begin with, few people were interested in radioactivity – the newly discovered x-rays were the centre of attraction. But as the novelty of these wore off, scientists began to work out a theory of what radioactivity was all about.

One of the first discoveries made was that radioactivity involved the gradual loss of mass – or 'decay' – of the material involved. It then emerged that it was the nuclei of the atoms themselves that were decaying. In the process, something strange occurred: electrons could sometimes be detected emerging from the debris. But weren't electrons supposed to be in orbit about the nucleus, not lurking *inside* it as well?

In the 1930s the Italian physicist Enrico Fermi (after whom the fermion family of particles is named) put forward an answer. The neutrons in the nucleus, he claimed, were being forced to turn into other particles: electrons, protons and antineutrinos. His calculations showed that the force involved was far weaker than the electromagnetic interaction holding the electrons in orbit about the protons. Fermi's force was thus dubbed the *weak nuclear interaction*. His idea of changling particles won strong backing from the discovery that when neutrons are isolated in the laboratory, they too disintegrate and give exactly the same three particles. It seemed clear that a new force of Nature had been uncovered.

The first glimmerings of unity

By the mid-1930s, Yukawa's radical new view of fundamental forces as the result of 'messenger' particles flitting from place to place was attracting the interest of many theorists. For the strong nuclear interaction, Yukawa had predicted the existence of the pi-meson. The obvious question now was which particle was responsible for the weak nuclear force? In 1938, Yukawa reran his calculations, arriving at a prediction of the properties of a new 'messenger' particle for the weak force, called, sensibly enough, the W-particle.

It turned out that, like the strong force, the newly discovered force was transmitted by a particle with a whole-number amount of spin – that is, by a boson. Intriguingly, the messenger particle of the electromagnetic interaction was also a boson – namely, the photon. Was this link between force carriers and bosons a coincidence?

On the face of it, this was very much a shot in the dark. Could things like light beams really be related to radioactivity? It seemed unlikely. The weak force was hardly like the electromagnetic force at all – for a start, it is a billion times weaker. For another, it had only a very limited range – affecting only subatomic processes. Electromagnetism, however, spreads out to the very edges of the universe.

This unpromising prognosis was not, however, enough to dispel the theorists' conviction that there was a deep unity in the universe. The links may not be obvious, ran the argument, but that does not mean they do not exist. By the mid-1950s, Julian Schwinger (who had helped iron out the problems of QED) had drafted a theory which appeared to bring the weak and electromagnetic interactions together.

It wasn't very convincing. It emerged that another particle was needed to weld the theories of the weak and electromagnetic interactions together. In 1961, the 29-year-old theorist, Sheldon Glashow, at the University of California at Berkeley, published the theory for this extra particle, now called the Z^0 (Zed-nought). But this breakthrough, in turn, caused problems. Glashow's theory seemed to push the analogy between the two forces too far. It implied that, like the photon responsible for the electromagnetic interaction, the W and the Z^0 had no mass. But the limited range of the weak interaction strongly suggested that the W and Z^0 must have mass – indeed, by particle standards, they should be heavyweights. So huge a disparity between theory and reality was hardly encouraging.

It says much for the strength of their belief in unity that the particle physicists did not reach the obvious conclusion from all this: that their idea of a deep connection between electromagnetism and the weak force was just wrong. Among those who held the faith were a schoolmate – and fellow theorist – of Glashow's, Steven Weinberg at Harvard, and the Pakistani Abdus Salam of Imperial College, London.

Working independently, they attacked the problem of the lightweight W and Z^0 using an idea put forward by Peter Higgs at Edinburgh University. Roughly speaking, Higgs showed that particles could become imbued with mass by 'eating' a particle, now known as the Higgs vector boson or simply the Higgs particle. The W and the Z^0 can both 'eat' a Higgs particle, thus getting mass, but the photon does not – precisely as observed in Nature.

And thus Weinberg and Salam proved that theorists had been right to hold faith with the idea of unity – the weak force and electromagnetic force really are different aspects of a single – 'electroweak' – force.

The reaction of the rest of the scientific world to this triumph of human creativity was silence. Weinberg and Salam's work was largely ignored when it was published. There were two basic reasons for this. Firstly, some experimental work done in the 1960s gave data that cast doubt on the new approach – but these data were later shown to be faulty. Rather more serious was the fear that the new 'electroweak' theory would prove to have a terminal case of the dreaded infinity disease.

With quantum electrodynamics, the mathematical trick of renormalisation had killed off the infinities. Many theorists decided that the new theory was not worth worrying about until someone had shown that it, too, could be cured of infinities.

The answer came in the form of numbers pouring out of a computer at the University of Utrecht, Holland, in 1971. A 25-year-old theorist called Gerhard 't Hooft had asked the computer to search for infinities in Weinberg and Salam's theory in all its different configurations. The hope was that for every infinity lurking in the theory, another one would pop up to cancel it out.

Did each infinity really have an obliging partner? Having had to pile so many extra concept on to what had started life as a simple

Re-creating the earliest moments of the universe:
the Large Electron–Positron Collider (LEP) at CERN, Geneva

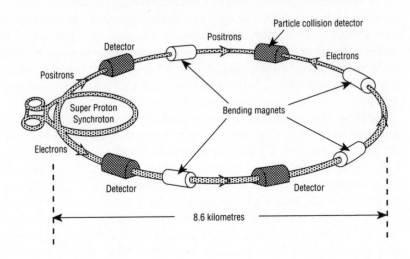

idea, it seemed almost too much to hope that the final mass of mathematics could be a reflection of the mind of God. 'T Hooft recalls the almost miraculous sight that greeted him that day in 1971: 'The output of the program was an uninterrupted string of zeros. Every infinity cancelled exactly.'

This astonishing result had an electrifying effect on the physics community. Plans were drawn up to find evidence for the existence of the W and Z^0 in the world's particle accelerators. In 1973, the first bits of evidence came in: so-called 'neutral currents' were found by researchers at CERN. Plugging data from this discovery into the electroweak theory enabled the masses of the W and the Z^0 particles to be worked out. They turned out to be huge: 87 times heavier than the proton for the W, and 98 times heavier than the proton for the Z^0. At least this meant that no one could be accused of missing the particles in past experiments: the machines then being used simply weren't powerful enough.

The proof of the unity

In 1976, Carlo Rubbia, an Italian experimental physicist at CERN, saw a way out. If he sent protons round his laboratory's two-kilo-metre wide Super Proton Synchroton machine in one direction, and antiprotons round in the other, they would crash into each other with sufficient energy to create the W and Z^0 – if they existed. The conversion of the SPS took three years to complete, and was masterminded by a Dutch engineer, Simon Van der Meer.

In 1981, the machine was switched on and the first protons and antiprotons hurtled round the SPS in the search for the W and Z^0. By the end of 1982, more than a billion proton and antiproton collisions had taken place in the machine. Yet only a handful were expected to lead to the creation of the heavy Ws and Z^0s. Picking them out would clearly require great skill.

The CERN experimentalists found a way of picking out just the one in 1,000 collisions in the machine that was worth even recording. But that still left data for the remaining one million collisions to be scoured. Yet, by meticulous searching, the CERN team was able to boil the million recorded collisions down to just 39 events that may just harbour evidence for the W and Z^0. Using a special computerised imaging system to inspect these events for telltale signs of the W and Z^0, sixteen 'finalists' were selected.

Five of these showed signs of having large amounts of energy missing – as if some particle had been created and then flown off, out of the machine. Here, at last, was evidence of the W. Out of the million collisions recorded, just five had involved the creation of a W particle. Its mass, moreover, was just as predicted by Weinberg and Salam's theory. The following year, the intensity of the machine was stepped up to find the heavier Z^0 particle. It too eventually showed itself and had the mass predicted by the theory.

For their painstaking work in finding the W and Z^0, Rubbia and Van der Meer shared the 1984 Nobel prize for physics. The theorists Weinberg, Salam and Glashow had already received theirs, five years earlier. The Nobel committee had decided that the cleverness of the theory and the discovery of the neutral currents in 1973 made the finding of the W and Z^0 almost a formality. Even so, Salam admits to still being astonished by the way it all worked out so well in the end: 'It is always incredible that what people work out actually does happen.'

In 1989, the discovery of the Z^0 particle paid a nice extra dividend: it showed that the fundamental building blocks of matter really are just twelve in number – the same figure theorists had finally reached in their efforts to categorise the subatomic particles years earlier.

The number emerges from the incredibly short lifetime of the Z^0: one ten million billion billionth (10^{-25}) of a second. The speed with which the Z^0 falls apart depends on the number of different ways it can do it – the more routes, the shorter the lifetime of the Z^0. Whichever route it chooses, it leaves behind other particles, including – crucially – neutrinos. So the faster the decay of the Z^0, the greater the number of types of neutrinos that exist in Nature. In October 1989, the results of indirect measurements of this incredibly short life span of the Z^0 were announced by independent teams at CERN in Geneva and SLAC at Stanford. They showed that there are only three types of neutrino – as expected by the theorists. Three types of neutrino, with their lepton partners, the electron, muon and taon, makes six leptons in all. And, as each lepton has a quark partner, the CERN and SLAC results confirm that just twelve different types of particle are needed to build all the matter in the universe.

The most important implication of the discovery of the W and Z^0 remains that there are, in fact, not four but three fundamental forces in the universe: gravity, the strong nuclear force and now the 'electroweak' force. Yet can this be the end of the unification?

The grand unification that isn't so grand

With gravity being – as we shall see – so horrendously difficult to deal with in the language of quantum theory, it was somewhat inevitable that it would be the strong nuclear force that would be the next to yield to the drive for unification.

Oddly enough, the ideas behind it emerged from an attempt to rid the Weinberg and Salam theory of a problem it didn't have. In 1971, experimentalists had looked for the so-called neutral currents predicted by this theory but had seen nothing. The neutral currents eventually turned up in 1973. But at the time, Glashow and his colleague Howard Georgi thought it might be worth searching for other ways of unifying the weak and electromagnetic forces.

They were seeking a neater theory than Weinberg and Salam's, too. For a start, Weinberg and Salam's theory failed to shed any light on such matters as why the charge on the electron is the basic unit of charge for particles. Its equations also contain what can be described as 'fudge factors': certain quantities that can be given more or less any value one likes, thus guaranteeing agreement with observation. Physicists aren't impressed with such fudge factors, and the more of them that occur in a theory, the less impressed they are likely to be.

During the early 1970s, Glashow and Georgi churned out many varieties of electroweak theory. They all went wrong for one reason or another. One day in 1973, Glashow had an apparently crazy suggestion for his colleague: perhaps they ought to try and unify the strong force as well.

This did not seem very sensible – if there were problems unifying just two forces, bringing in another could only be worse. Georgi recalls the day Glashow made his wild proposal: 'Neither of us had the faintest idea how to find such a unification. We talked inconclusively until it was time to go home. After dinner, I retired to my living room, stretched out in a comfortable reclining chair and tried to think about the question more systematically.'

Incredibly, the introduction of the strong force seemed to make everything work. Working long into the night, Georgi succeeded in producing a theory that unified the electromagnetic, weak *and* strong forces. The result is now known, for technical reasons, as 'minimal SU(5)'.

As the hours slipped by, Georgi became increasingly excited with the way in which SU(5) appeared to explain so many mysteries. For example, it provided a natural explanation for the characteristic electric charges on all particles, even the quarks with their fractional charges.

But in the early hours of the next morning, Georgi made a disturbing discovery. His theory predicted that all atoms must eventually decay. Specifically, SU(5) predicts that protons, the particles that hold electrons in their orbits in atoms, break down into a neutrally charged pion and a positively charged positron. Up until this point, Georgi had been very happy. Now he became very worried. Hydrogen, which consists of just a proton and an electron, has been around since the birth of the universe, so if the proton does decay, it must do so very slowly. But Georgi could think of no way of forcing his equations to make this decay slow enough.

Essentially, the problem was that the theory demanded the existence of a new particle, which Georgi named the X-particle. Usually particles either have no mass at all, or a mass comparable to that of the proton, give or take a few powers of ten or so. But a slow rate of decay for the proton means that the X-particle must be far heavier than the proton. Georgi couldn't see a way around this problem. He therefore took the sensible course of action: he went to bed.

The following morning, he ran through his work with Glashow, explaining how everything, apart from the decay of the proton, seemed to fit. Incredibly, Glashow wasn't in the least bit concerned, recalls Georgi. 'Instead, he dragged me up to the library where we looked up the most recent data on the lifetime of the proton. It was said to be greater than 10^{29} years. That is a long time! Undaunted, Shelley estimated that if the X-particle were 10^{14} times as heavy as a proton or more, the proton would live long enough.'

Georgi still found this hard to take; such a mass was about a billion times greater than anything considered before by physicists. Even so, both he and Glashow were impressed by the mathematical beauty and neatness of SU(5) and decided to stick with it. They wrote up their work and got it published. It generated some amusement among their colleagues, many of whom thought the pair had finally lost their marbles.

The following year, working with Steven Weinberg and Helen Quinn, Georgi answered an obvious question about the unification he was claiming to have brought about. How can three different forces, with radically different strengths, all be aspects of a single force? The answer lies in the fact that the strength of the three forces varies with distance. In the world we normally see, the strengths of the three are radically different. But as one gets down to the scale of the atomic nucleus and smaller still, the strengths start to become comparable to one another. Finally, Georgi and his colleagues calculated, at the unimaginably small scale of one ten thousand billion billion billionth of a metre (10^{-31}m), the strengths of the three forces become the same.

By 1977, SU(5) was attracting a lot more interest, and had even been graced with the title of 'grand unified theory' or GUT – a rather misleading title given the fact that, at best, it unifies only three of the four fundamental forces; gravity is still left out. But there was something of a problem facing anyone hoping to win a Nobel prize by finding direct evidence for the truth of SU(5). It seems certain that no one will ever be able to build a particle accelerator capable of reaching the energies needed to create the incredibly heavy X-particles.

Fortunately there is a less direct route of testing SU(5): its prediction that all atoms fall to pieces if you wait long enough – or that, as Glashow put it, 'Diamonds aren't forever'. Using the simplest version of SU(5), Georgi calculated that the protons should typically live for about 10^{29} years. This is more than a billion billion times longer than the age of the universe, so how on Earth can such a prediction be tested? The solution emerges by rewording it: in any year, a proton has one in 10^{29} probability of falling apart. So, to stand a chance of seeing one proton falling apart in a year, all you have to do is watch 10^{29} of them. This might sound a lot until one realises that just a gram of the gas hydrogen contains about 10^{24} protons. Thus just a tonne of matter contains about 10^{30} protons.

Experimentalists seized on this idea and work began on proton-decay detector experiments in America, Europe, Japan and Africa. To insure themselves against the possibility that the theory's prediction is a little inaccurate, the experimentalists built their experiments far larger than SU(5) demands. The largest has been built about 600 metres underground in the Morton salt mine near Lake Erie, Ohio, by a team from the University of California at Irvine, the University of Michigan and the Brookhaven National Laboratory. The 'IMB' detector consists of a huge cubic cavern cut out of salt and filled with 8,000 tonnes of ultrapure water. Around the edge of the cube are set more than 2,000 light detectors. These are designed to pick up the telltale flashes of light caused when the particles created by a proton decay pass through the water.

The good news so far is that every detector has picked up these flashes of light. The bad news is that all the flashes could have been caused by cosmic rays. By putting their detectors far underground, the experimentalists have tried to minimise the problem of spurious signals from cosmic rays, but some still get through.

The current consensus is that proton decay has not yet been seen. The size of the detectors, when taken with the amount of time they have been running, thus implies that the proton lives for at least 10^{32} years – a thousand times longer than the simplest form of SU(5) predicts. This doesn't mean that SU(5) is wrong – just that its simplest version seems to be inaccurate, which is hardly surprising.

The impossible magnet

Other teams of experimentalists are searching for confirmation of another prediction of SU(5): that a weird type of magnet with just one pole – a monopole – exists somewhere in the universe.

Monopoles are not simply tiny, bar magnets cut in half. If you cut an ordinary magnet in two, you just end up with two smaller magnets, each with a north and a south pole as before. Monopoles emerge from SU(5)'s explanation of the otherwise completely mysterious fact that electric charge always comes in discrete packages. Why don't some particles have an electric charge of, say, 0.6854323 times that on the electron? As early as 1931, the British physicist Paul Dirac showed that this 'charge quantisation' meant that monopoles must exist. In 1974, the Dutch theorist 't Hooft and the Russian Alexander Polyakov independently showed that charge quantisation in SU(5) – indeed, in any unified theory of a similar type – also implied that monopoles must exist.

So how can they be detected? Being creatures spawned of the incredibly high energies of GUTs, these monopoles are expected to have masses comparable to that of the X-particle, some 10^{16} times that of the proton. As such, monopoles would be more or less unstoppable by ordinary matter: none of the four forces is strong enough to halt them as they pass by. So scouring the Earth for them is unlikely to uncover their existence. A better approach is to try and detect them as they fly past a suitable detector.

In 1982, Blas Cabrera at Stanford University in California started electronically fishing for monopoles. His 'net' consisted of a coil of metal five centimetres across through which an electric current flowed. If a monopole passed through, it would cause a sudden jump in the electric current and a squiggle on the recording-equipment paper.

Just before two o'clock on the afternoon of St Valentine's Day 1982, Cabrera picked up just such a pulse. This event caused great excitement in the world of physics. But, like the proton-decay experiments, the consensus is that the Cabrera result was a freak; in 1990, Cabrera himself said it should now be 'discarded'. Other, much more sensitive detectors have since been built, but apart from an unconfirmed monopole event in 1985 at Imperial College, London, nothing has been seen so far.

Despite this somewhat depressing dearth of hard evidence supporting the 'grand unification' of three of the four forces, confidence that the theorists have at least got the right approach remains high. This is essentially because of the neatness of the idea – particularly its explanation of the quantisation of charge. Theorists think it 'smells right'.

But there is still plenty that is unsatisfactory. For a start, Georgi and Glashow's 'minimal SU(5)' is not the only possible GUT; it's just the simplest. The proton decay experiments, as we have seen, suggest that it needs to be modified, but how? Theorists have found ways of constructing other candidate GUTs, all broadly similar to minimal SU(5), and it is not clear how to choose between them. Things are not helped by the fact that the equations of all of them contain fudge factors.

The numbers of fudge factors in particle physics has become something of an embarrassment. John Ellis, head of theory at CERN in Geneva, recently totted up the number of them in the so-called Standard Model – that is, the combined theories of all forms of matter and the forces between them. He counted twenty: a ragbag of masses of quarks and leptons, force strengths, and interaction parameters. The majority of them are, however, related to the mass of the Higgs particle – the particle that Weinberg and Salam called in to give the W and Z^0 particles of the electroweak force theory some mass. Finding the Higgs particle would thus do more than put the icing on the cake of the electroweak theory. It would also make the Standard Model look a lot more convincing. Of course, if the Higgs turns out to have some outlandish mass, it could also wreck the Standard Model. Small wonder that the Higgs, along with the Top quark, heads the 'most wanted' list of the particle accelerator teams.

But what theorists would dearly like to find is some *principle* that forces these fudge factors to take on certain values, thus ridding the equations of their arbitrariness. After all, one would have thought that there can be only one correct theory for GUTs.

Perhaps Glashow's 'crazy' idea of solving the problems of unification by bringing in another force could be made to work again. Perhaps the problems of SU(5) can be answered by bringing in the fourth and final fundamental force: gravity.

Thus far we have studiously ignored gravity. On the face of it, this omission seems rather odd; none of the four fundamental forces is more familiar than gravity. But it turns out that there are profound problems in bringing gravity into the fold. Its familiarity belies a host of unpleasant surprises for anyone trying to incorporate it into a *truly* grand unified theory bringing together all four forces – a 'Theory of Everything'.

From an apple garden to the cosmos

By a long, long way, gravity is the most feeble of all the four fundamental forces. Even the so-called weak nuclear force is 10^{32} times stronger. Yet gravity is the most important force in the universe. While the other forces either affect only a certain type of particle, or have very limited ranges, or can be reversed, gravity affects everything, everywhere.

The conception of the *universality* of gravity was the crowning achievement of the genius of Isaac Newton. And, for once, the old story about the origin of that brilliant idea seems to be more or less true. The 24-year-old Newton was sitting in the garden of his mother's house in Lincolnshire one day in 1666 when an apple fell from a tree. It was this that set him thinking about the power by which one object might attract another. A contemporary biographer who interviewed Newton years later then learned how Newton's genius has extended the idea beyond a country garden and into space. 'As this power is not found sensibly diminished at the remotest distance from the centre of the earth, to which we can rise, neither at the tops of the loftiest buildings, nor even on the summit of the highest mountains, it appeared to him reasonable to conclude that this power must extend much farther than was usually thought; why not as far as the moon, he said to himself? and if so, her motion must be influenced by it; perhaps she is retained in her orbit thereby.'

To make this breathtaking leap from a falling apple to the motion of the moon, from Earthbound phenomena like the fall of an apple to a law which applies throughout the universe, takes the type of creative arrogance that marks out the greatest scientists. Using astronomical observations of the planets, Newton formulated a law which stated that the force of gravity between two objects is proportional to the product of their masses (i.e. one mass multiplied by another) and to the inverse square of the distance between them.

Armed with this discovery, Newton tried to find proof for his idea. If this force of gravity were truly ubiquitous, it should, he reasoned, be possible to work out the strength of the force needed to hold the moon in its orbit from a knowledge of the strength of gravity at the Earth's surface. In the seventeenth century, the distance of the moon was known to be about 60 Earth-radii (the modern value is 60.27 Earth-radii). But Newton's calculations demanded that the moon's distance be expressed in miles; this meant that Newton needed the radius of the Earth in miles as well. Unfortunately for Newton, this was thought to be about 3,500 miles – ten per cent less than the real value. So, when Newton plugged the values into his formula for gravity, he found that his results were only 'pretty nearly' the values he was looking for. This, for Newton, was not good enough. For some reason, he did not stop to question the accuracy of the radius measurements. Perhaps, he wondered, some other force as well as gravity was at work? Disappointed by his near miss, Newton 'threw aside the Paper of his Calculation, and went on to other studies'.

Not until 1675 did Newton learn of a remeasurement of the radius of the Earth – to a value much closer to that which we now know to be correct. Putting the new value into his long-neglected law, Newton found that the answers were more or less exactly as they should be. He had, it seemed, been granted a very fleeting glimpse into the mind of God.

Newton went on to apply his law of gravity to the rest of the solar system and eventually to the whole universe. His followers, notably his friend Edmund Halley, used the law to predict new phenomena such as the return of the comet which bears Halley's name. No longer was the universe a capricious place, where evil comets arrived to presage some impending disaster – its finest details could, it seemed, be foretold by Newton and his followers into the infinite future.

In the 1840s, using Newton's law to study wobbles in the orbit of the planet Uranus, a 26-year-old Cambridge mathematician named John Couch Adams and the French astronomer Urbain Leverrier independently predicted the existence of a new planet, Neptune, later discovered in 1846.

This discovery was seen as the greatest triumph of Newton's theory. But ironically, just thirteen years later, Leverrier himself was reporting to the Academy of Sciences in Paris a discovery which was to be the death knell for Newton's conception of gravity.

The 'notorious' general theory of relativity

The realisation that something was wrong with Newton's theory emerged from some amazingly meticulous measurements made of the orbit of Mercury, the planet nearest to the sun.

As a result of the gravitational attraction of the other planets, the entire orbit of Mercury is slowly hauled around the sun. It is possible to calculate just how fast the orbit is thus being moved around. What worried astronomers was that Newton's law of gravity could account for only about 99.5 per cent of the total amount of rotation; there was a tiny bit left over. The discrepancy really was tiny: equivalent to an extra three ten-millionths of a complete rotation every year.

Before the discovery of the planet Neptune, some astronomers had thought that Newton's law might have to be altered slightly as one went deeper into space. But by the time of Leverrier's announcement about Mercury, the consensus had changed: Newton's law must be right, and this new discovery must hint at the existence of some new source of gravitational influence.

Leverrier thought the explanation might be a new planet orbiting even closer to the sun than Mercury. It was given the name Vulcan, and some reputable astronomers even claimed to see it during eclipses of the sun. Most astronomers remained unconvinced, however, and declared the mystery of Mercury's orbit unsolved.

It remained that way for over half a century. Then, in 1915, Albert Einstein showed that the Mercury mystery was no mere mote in the eye of science. It was a veritable beam, and it demanded a whole new theory of gravity. That theory is called the general theory of relativity, or just General Relativity (GR). It gets its name from being essentially an extension of the 'special' theory of relativity, published by Einstein ten years earlier. It is also notorious for its mathematical complexity.

Special relativity deals with how fundamental concepts such as length and time are affected by high speeds, particularly speeds close to that of light. It allows one to calculate things such as the relative velocities of two subatomic particles racing towards each other in an accelerator.

The theory basically deals with measurements made in situations where motion is at a steady velocity – general relativity, as its name suggests, is more general. It concentrates on objects that are

Einstein's great idea: how to mimic the effect of gravity in a rocket

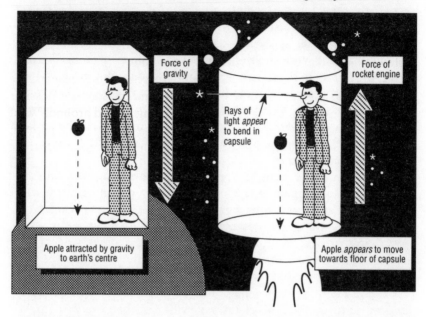

accelerating relative to one another. In particular, it deals with that most familiar source of acceleration in objects: the force of *gravity*.

Ironically, the fact his special theory of relativity could not handle gravity dawned on Einstein while he was writing a review on what special relativity *could* do. He determined to find a way of extending his theory. For some months, no obvious way forward presented itself. Clearly, however, his subconscious was churning away on the problem. For, a few months later, he was sitting in a chair in the patent office in Bern, Switzerland, when he had what he later described as the happiest thought of his life. 'All of a sudden a thought occurred to me: "If a person falls freely he will not feel his own weight." I was startled. This simple thought made a deep impression on me. It impelled me towards a theory of gravitation.'

The reason was that it provided a link between the effects of gravity and the effects of acceleration caused by other means. This equivalence may not be obvious, so let us use the technique of which Einstein was so fond: the thought experiment (see diagrams above).

Picture a spaceship in deep space, far from any stars, planets or other sources of strong gravitational forces. Inside, our astronaut is just floating around, completely weightless – there is no force acting to make him do anything else. Suddenly, the on-board computers decide it's time for the engines to come on and send the rocket forward. Seen from outside the rocket, the effect of the ignition is to make the floor of the spaceship rush up under the free-floating astronaut who, along with anything else that happened to be floating about, thus hits the floor with a resounding crash. But inside, where every floating object has hit the floor, the effect of the rocket engines is remarkably similar to what would have happened if someone had switched on a downward-acting gravitational field inside the rocket. This similarity between gravity and acceleration Einstein called the 'principle of equivalence'.

Even in this crude form, the principle has some surprising predictions to make. Imagine that as the rocket is powering forward a passing alien spaceship fires a laser beam through one window of the capsule. As the beam crosses the capsule, the rocket continues to power upward. As a result, the beam hits the other side of the capsule slightly below the point at which it came in. That's not very surprising. But something more surprising emerges when we use our new-found equivalence principle. From this we know that whatever happens with upward accelerations will also happen with gravitational forces. So we can make a prediction: gravitational fields can bend rays of light.

This all seems rather simple. So why did GR acquire its reputation for appalling difficulty? This comes from the fact that the principle of equivalence works only up to a point. It is always possible to find slight differences between accelerations and, say, the gravitational effect of a planet. For example, in our thought experiment, if you drop two objects in the spacecraft when the engines are propelling it, they will fall to the floor in two exactly parallel paths. But drop the two objects when the rocket is sitting on a planet experiencing the pull of gravity, and their paths will *not* be parallel – they will both be pointing towards the centre of the Earth, and thus will be bent towards one another.

Einstein thus had to take into account the limitations of his principle of equivalence; he also decided to build in some other, more technical, features to ensure general applicability of his final equations. Finally, Einstein had to be certain of keeping all of the successes of Newton's law of gravity. This meant that when gravitational fields are quite weak and velocities are much below the speed of light, the new theory had to boil down to Newton's old law.

What Einstein means by gravity

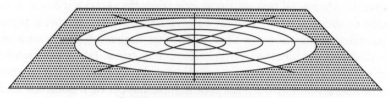

Spacetime without a mass within it is flat –
ordinary geometry applies

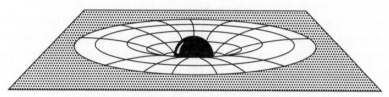

Mass *curves* spacetime, rather like a cannonball in a rubber sheet.
This curvature, according to Einstein, constitutes 'gravity'

To do all this, Einstein had to use some very sophisticated mathematics that had only recently been developed. It is this 'tensor calculus' that strikes fear into the heart of many students today.

However, we do not need all the mathematics to understand the final picture of gravity that emerges from GR. It is that gravity is not, as we so often view it, some sort of elastic stretching from, say, the Earth to an apple in Newton's garden. According to Einstein, gravity is essentially *a curvature of space and time.*

With this new basis, the Earth can be pictured (very approximately) as some sort of huge cannonball sitting in a kind of rubber sheet of space and time ('spacetime'), and curving both (see diagram above). Newton's apple falls off the tree, and follows a path down the rubber sheet towards the centre of the Earth.

Einstein's theory is put to the test

As early as 1907 Einstein had set himself the goal of explaining the mystery of Mercury's orbit with his emerging theory of gravity. By November 1915 he had reached the stage where his equations could be used to investigate the mystery. The theory leads to a formula for the extra amount of movement that cannot be accounted for by Newton's theory. When he fed in the necessary data, the result was a rate of movement for the orbit of Mercury of three ten-millionths of a complete rotation per year – precisely the amount measured by the astronomers all those years earlier. 'For a few days I was beside myself with joyous excitement,' he later wrote. He told a friend that the excitement was so intense that his heart went into spasms.

The way in which the equations were able to explain this years-old mystery with no jiggery-pokery or fudge factors was enough to convince Einstein that GR was indeed the successor to Newton's law. But it was his successful prediction of another effect of gravity that turned Einstein from a brilliant scientist into the apotheosis of genius.

The principle of equivalence, as we have seen, leads directly to the idea that gravity can bend light. In fact, Newton's old theory also predicts light bending; the German astronomer Johann Soldner calculated the tiny effect as far back as 1801. But as with Mercury's orbit, GR gives different answers when the numbers are put in. Or at least, it should; Einstein came within an ace of making a dreadful mistake when he used an early version of his theory to calculate the bending of starlight by the sun.

During his search for the final equations of general relativity, and even before he realised that gravity meant curvature of space, Einstein had made a stab at working out the amount of bending, expressed as the angle by which the light is deflected from a straight line. He obtained a figure of 0.87 seconds of arc. This is a tiny angle, equivalent to the visible size of a needle seen from a distance of about 250 metres.

Einstein at first feared that it might never be possible to measure so small a deviation. However, he came up with a suitable test which exploits a curious fact about the Earth and the moon. Of all the planets in the solar system, the Earth is the only one whose natural satellite appears almost exactly the same size in the sky as the sun. This means that, during a total eclipse of the sun when the moon comes between the sun and the Earth the sun's light is cut off for up to seven and a half minutes. This allows even those stars lying next to the sun in the sky to come into view.

By measuring the positions of the stars very accurately both before and after the eclipse, it is therefore possible to detect any change caused by the presence of the sun and its gravity field in their part of the sky.

Einstein was held in sufficiently high regard by the scientific community to have his ideas about light bending taken very seriously. Thus, from 1912 onward a number of expeditions set off to look for signs of light bending during total eclipses.

Fortunately for Einstein, they were all called off for one reason or another. If they had measured the light deflection, they would have found Einstein to be wrong. Bad weather and war at the eclipse sites actually saved Einstein from huge embarrassment.

By November 1915, Einstein had realised that gravity is the result of the curvature of spacetime, and was able to recalculate the bending of light by the sun. This time he got an answer of 1.75 seconds of arc, twice the original, flat-space value. It was also twice the value calculated by Soldner on the basis of Newtonian gravity – and thus a test of the new theory's greater powers.

An expedition to view an eclipse from Venezuela the following year had to be called off because of World War I. Attempts to use old photographs of stars during past eclipses also failed. In the end Einstein had to wait four years before his theory could be put to the test. It followed the arrival of the first papers on GR to reach England. They had been sent to the Cambridge astrophysicist Arthur Eddington, who immediately urged the Royal Astronomical Society to set up an eclipse expedition. The Astronomer Royal, Sir Frank Dyson, realised that the eclipse of 29 May 1919 could hardly be bettered as a time for testing Einstein's prediction. The sun would be passing through a part of the sky containing a star cluster, packed with stars whose light would skim the disc of the sun during the eclipse.

To reduce the risk of a washout, two separate expeditions were arranged, one to go to Sobral in Brazil, the other – led by Eddington – to the island of Principe off the West African coast. On the day of the eclipse, Sobral had perfect skies. Principe, however, had overcast skies which threatened to ruin the experiment. But at the crucial moment, the clouds parted and both sites were able to make their observations.

There followed months of careful analysis of photographic plates taken before and during the eclipse. In September 1919, Eddington coyly revealed to a meeting of the British Association for the Advancement of Science that the bending of light found was somewhere between Einstein's original value and his revised value. Tension mounted. Finally, at the end of September, Einstein received a telegram from a friend, the Dutch physicist Hendricks Lorentz. General relativity had won the day. The deflection found by the two expeditions averaged 1.8 seconds of arc, double the value predicted by Newton's old theory but in virtually perfect agreement with Einstein's prediction.

On 6 November 1919 a joint meeting of the Royal Society and the Royal Astronomical Society was called formally to announce the confirmation of Einstein's theory. According to Einstein's biographer Abraham Pais, this was 'the day Einstein was canonised'. By the following morning he had become a living legend. *The Times* in London carried successive articles on the 'revolution in science' which had seen 'Newtonian ideas overthrown'. The national newspapers of other countries followed. By 18 November, the legendary difficulty of GR had become established, with *The Times* asking its readers not to take umbrage over the fact that 'only twelve people in the world' could understand the work of 'the suddenly famous Dr Einstein'.

Since those days, nothing has diminished the brilliance of Einstein's theory of gravity. Other, more sensitive and esoteric, tests have been devised; all have been passed with flying colours. From time to time, rival theories of gravity emerge to challenge GR. To date, they have all either failed some crucial test or have been shown to boil down to Einstein's theory when compared with observational data.

Claims are sometimes made that the match between Einstein's predictions and observation is just too good. For example, it has long been known that if the sun is slightly fatter at its equator than at its poles, its gravitational field would provide an additional push to the orbit of Mercury. If the sun has a big enough equatorial bulge, it would spoil the close agreement between the observed rate of movement of Mercury's orbit and the extra motion predicted using GR. The consensus, although some still disagree, is that the sun's bulge – if it exists at all – is not big enough to invalidate the theory.

Confident from his successes within the solar system, Einstein went on to apply his theory to the universe as a whole. We will look at the results in the next chapter, on the origin and fate of the universe. Suffice it to say now that the results support the view that GR is the best theory of gravity we have.

Yet for all these successes, an outstanding question remains: how can we square Einstein's view of gravity with what we know about the other fundamental forces of Nature? In attempting to answer this, we begin our investigation of what is arguably *the* outstanding mystery of modern physics: will there ever be a 'Theory of Everything'?

THE OUTSTANDING MYSTERY

Will there ever be a theory of everything?

Theorists believe that there is a fundamental unity underlying the fundamental forces of Nature. If they are right (and they may not be), then gravity is, at bottom, just one more facet of a single 'super-force', of which the strong, weak and electromagnetic forces are also just a part.

But until we met gravity, we had viewed all the fundamental forces as being due to the exchange of messenger particles, such as the photon for the electromagnetic interaction and the gluons for the strong force. Yet according to Einstein, gravity is a manifestation of something completely different: curvature of space and time. How are we to marry these two disparate views of 'force' together?

Using Einstein's equations, it is possible to show that for gravity there is a messenger particle called the graviton. Like all the other messenger particles, the graviton is a boson. Like the photon, it is also massless – a reflection of the fact that, like the electromagnetic force, gravity has an infinite range.

But as soon as we start to treat gravity as just another quantum field like electromagnetism, we run into trouble. We get a severe case of the infinity disease and this time the usual palliative – renormalisation – fails to work. Some theorists have taken this to be a warning that gravity is *not* just another quantum field, and that we must now be a lot more careful how we proceed. Even the non-expert can see the sense of this: all the other quantum field theories, such as QED, can blandly talk about things happening in space and time without fretting about some nasty link between the quantum field and space-time itself. But if Einstein told us one thing, it is that gravity is *all about* spacetime – and so surely, when quantising gravity, there is at least a strong possibility of some nasty link emerging?

There is a bit of a problem in raising such questions, however. Grappling with them is extraordinarily difficult mathematically, even by the extremely high standards of theoretical physics. Although the last few years have shown signs of a breakthrough with these problems, trying to create quantum gravity out of GR alone remains the tough guy's option.

What a Theory of Everything must bring together

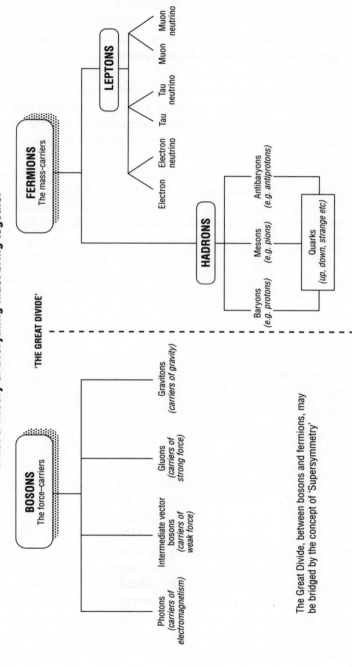

The Great Divide, between bosons and fermions, may be bridged by the concept of 'Supersymmetry'

Rather than spend years staring at what may well be insoluble equations, the bulk of theorists have decided to plough on with the idea that gravity really is just another quantum field, and that one day a Glashow or a Salam or a Weinberg will find a way through.

Even so, unifying gravity with the other forces has long seemed a task beyond even the greatest minds. Einstein spent the last three decades of his life fruitlessly trying to unify just gravity and electro-magnetism, the only other fundamental force known when he started. The day before he died, in April 1955, he was still trying.

But now, in the 1990s, there is a buzz of excitement among theor-ists that the dream of a unified theory may at last be realised. Ironic-ally, it seems that at one stage Einstein may have been on the right track, although the gap between his work and the theory now being investigated is immense – not surprisingly, given Einstein's rejections of the basis of conventional quantum theory.

Descent into the fifth dimension

The link comes from a letter Einstein received in 1919 from a German mathematician named Theodore Kaluza, which contained details of an intriguing discovery. It centred on the 'dimensionality of space'. Whether we realise it or not, we all conventionally think of the world as four-dimensional. In the language of physics, this means that to define precisely a particular phenomenon in the universe, we need four numbers. For example, suppose you want to tell a lost explorer to meet you at a hut on a mountainside. To give the explorer the details, you would need four numbers. Three give the precise location of the hut: longitude, latitude and height above sea level. The fourth number – the fourth dimension, that is – gives the precise *time* of the rendezvous.

Although four dimensions would seem enough for anyone, math-ematicians know no such bounds. If you want a universe with more dimensions, you just add on extra terms to the equations – there are no mathematical difficulties in thinking about a universe with a hundred dimensions, if you really need to.

Einstein's equations giving the spacetime curvature (i.e. gravity) produced by a given amount of matter are normally written for just four dimensions. What Kaluza found was that if GR was written out for a five-dimensional universe instead, you still got Einstein's theory, but you got something else as well: the laws of electromagnetism! Could it be that electromagnetism is merely a manifestation of gravity in some hidden, fifth dimension of the universe?

Kaluza appeared to have achieved a unification of gravity and electromagnetism by adding an extra dimension to the universe. He can thus be regarded as the father of unified field theories. Einstein was initially very enthusiastic about Kaluza's idea. But some questions clearly need answering. Do we really live in a five-dimensional universe? If so, where is this extra dimension?

By 1926, the Swedish mathematician Oskar Klein had found a clever explanation for why we don't see the curious fifth dimension: it is too small. According to Klein, it has been very tightly 'curled up' or, to use the modern jargon, 'compactified'. To help visualise this bizarre idea, picture a garden hose. Seen from a distance, the hose looks like a thin, one-dimensional line. Get closer, and one sees that the line has thickness – an extra dimension has come into view. Get closer still and the fact that the hose has a circular cross section, and is thus really three-dimensional, finally becomes clear.

Klein was even able to calculate just how tightly curled up this fifth dimension is. The equations of Kaluza's five-dimensional gravity show that the 'size' of the bundled-up fifth dimension depends on the amount of electric charge on the electron. Putting in the figures, one finds that the missing fifth dimension of the universe is packed into a bundle about 10^{-32} metres across. This is almost unimaginably small. To see an object of this size, one would have to use a magnification so great that an electron would be about the size of our solar system. No wonder, then, that we cannot see the fifth dimension of the universe.

Although Einstein later lost interest in the Kaluza-Klein theory, its intriguing ability to bring both electromagnetism and gravity under one roof continued to excite others for some years. But then came the discovery of two more fundamental forces – the strong and weak nuclear forces. It was far from obvious how to build these into the Kaluza-Klein theory. Perhaps, thought theorists, the theory was little more than a mathematical quirk after all.

Supersymmetry lights the way

For decades afterwards, gravity was neglected by the quantum theorists. Understandably enough, most directed their effort at areas where
they knew they could make some progress. The quantum theory of
the electromagnetic, strong and weak forces blossomed. Then attention turned to finding deep connections between these three forces.
This also proved a rich and rewarding field. Nobel prizes were being
awarded as confirmations of the theory came in.

At the heart of so many of the successes was a concept theorists
had discovered to be almost impossibly powerful: *symmetry*. In the
1960s it was most apparent in the work of Gell-Mann and others,
who, as we saw earlier, found it capable of bringing order to the
myriad collections of particles emerging from the world's particle
accelerators. The Eightfold Way was graphic evidence for the existence of symmetry at a fundamental level. Symmetry also manifested
itself in more subtle, abstract ways in the mathematics behind theories
proving connections between all the fundamental forces but gravity.
Was there a symmetry principle that would bring gravity into the fold?

In the early 1970s, theorists found an extremely powerful new
symmetry principle that rekindled hopes of attaining that holy grail
of physics. It is known, appropriately enough, as *supersymmetry*.
The excitement stemmed from the apparent ability of supersymmetry
to bridge the great divide of particle physics: those particles with
whole-number amounts of 'spin' – the bosons – and the others, called
fermions. Bosons and fermions are very different creatures; they obey
different rules, and play different roles in the universe. But most
crucially, all the messenger particles for the fundamental forces –
including gravity – are bosons, while all the particles of matter they
act on are fermions. Supersymmetry thus seemed to offer a way of
bringing together in one overarching principle *all* matter in the universe and *all* the fundamental forces that act on it. It looked as if
theorists had at last found a path to a Theory of Everything.

The effect on theorists of the discovery of supersymmetry was
dramatic. Around the world a veritable army of them set to work
seeing what supersymmetry would lead to. Something very exciting
emerged when it was applied to general relativity. The resulting
theory of 'supergravity' had very few fudge factors that could be
fiddled with to give the right results. This seemed a promising sign
– after all, anyone could build a theory which gave the right results

as long as there were enough fudge factors built in. A true theory of everything should leave no room for manoeuvre.

The most important bit of leeway supergravity *did* offer was in the number of a particular type of particle that could exist in the universe. There could be anywhere between one and eight different types. As theorists were hoping to get as much from supergravity as possible, they opted for the version which had the maximum number of these particles in it. And so was born the now notorious theory of N=8 supergravity.

At first, everything seemed very rosy. Any theory of everything must account for the existence of all the myriad particles discovered by particle accelerators. N=8 supergravity certainly gave a lot of particles. The supersymmetry principle predicts that there are a host of 'supersymmetric partners' for the known particles. For example, the boson that acts as the messenger particle in gravity – the graviton – is predicted to have eight fermionic partners known as gravitinos. The photon, gluon and W particles have partners, known as photinos, gluinos and winos. All the fermions have supersymmetric partners, too, such as selectrons and squarks. Clearly, N=8 supergravity was going to keep experimentalists happy for years to come, confirming the existence of all this lot. (In 1985, researchers at CERN thought they had found a squark, but they were wrong – and nothing has been seen since.)

But for theorists, the most exciting feature of supersymmetry was that a Theory of Everything based on it might be free of the infinity disease. Again, things looked pretty good to begin with. Some mathematicians discovered that a theory akin to N=8 supergravity was indeed totally free of infinities – it didn't even need to be treated for them with renormalisation. Better still, it seemed that all these particles in N=8 supergravity would lead to a magical cancelling out of any infinities that did pop up during calculations.

The trouble was, no one could face the appalling job of actually *proving* that supergravity was free of infinities. Professor Stephen Hawking of Cambridge, one of the earliest supersymmetry enthusiasts, explains the situation: 'It was suspected that some of the infinities might still remain. However, the calculations required to find out whether or not there were any infinities left uncancelled were so long and difficult that no one was prepared to undertake them. Even with a computer, it was reckoned it would take at least four years, and the chances were very high that one would make at least one mistake, probably more.'

Fretting about the infinities seemed to many theorists to be looking a gift horse in the mouth, however. Maybe someday some drudge would crank through the proof, they reasoned – in the meantime, there may be plenty of nuggets waiting to be dug out of this apparent goldmine of a theory. In 1979, a very big nugget was dug out: N=8 supergravity in our four-dimensional world was found to drop out of a simpler version of the theory known as N=1 supergravity. The condition was that the latter theory had to hold sway in an *eleven*-dimensional universe.

In other words, a comprehensive theory of everything in our own universe could emerge from a much simpler theory of everything in eleven dimensions. This renewed talk of universes with lots of dimensions sounded a lot like Kaluza's age-old idea. It wasn't quite the same – he had needed just one extra dimension to bring in electromagnetism. Supergravity required another seven, but then, it had also to account for two nuclear forces Kaluza never even knew about.

After so much wishful thinking about N=8 supergravity, this concrete result generated a huge boost in confidence among theorists. Caltech's Murray Gell-Mann went round the world telling everyone who would listen that supergravity was the theory to work on if you wanted the credit for building the Theory of Everything. In Cambridge, Stephen Hawking felt moved to announce that N=8 supergravity might mean that the end of theoretical physics was now in sight.

This now notorious pronouncement proved to be somewhat premature, to say the least.

Supergravity turns sour

The theorists now dug deeper into N=8 supergravity, looking for bigger and brighter nuggets. But soon all they were getting was dross. For a start, the plethora of particles the theory generated did not square with what was known about the quarks, leptons and the W particles. Its use of eleven dimensions also produced a lack of detail about an essential technical property of fermions; only by using an *even* number of dimensions did this problem disappear. Further investigation revealed that even this restriction on the number of dimensions was not enough. To get rid of the technical problems for both fermions and gravitons, only N=1 supergravity in ten dimensions or N=2 supergravity in six or ten dimensions seemed acceptable. By the early 1980s much of the neatness of the 1979 discovery had evaporated. Worse still, the two surviving versions were found to contain exactly the type of infinities everyone had been dreading at the outset. They also suffered from so-called 'anomalies' which made the theories inconsistent with our universe.

The small army of theorists that had started out with such high hopes for supergravity started to lose heart. Some new ideas were desperately needed. The breakthrough came one day in 1984. It had its origins in an idea that had long been regarded as an utter waste of time by all but a small group of renegade theorists. It went by the unprepossessing name of *string theory*.

The strings that bind the universe

String theory started life in the late 1960s, when the big problem was how to account for all these particles coming out of the accelerators. One particularly problematic family was the hadrons – the particles, such as the proton and neutron, that feel the strong nuclear force. While some were reasonably well behaved, certain hadrons had very short lifetimes, so short that they barely seemed to have an existence of their own. They were thought of as mere resonances of their more stable cousins.

Musing over the various strange properties of hadrons, an Italian theorist called Gabriele Veneziano guessed a formula which expressed many of the curious features of the hadrons. This prompted others to try to find a way of explaining why the formula worked. The idea that emerged was intriguing, to say the least. Like all particles, hadrons were usually thought of as pointlike objects. It turned out that Veneziano's formula could be deduced on the assumption that hadrons are, in fact, tiny *strings*. A neat image of the hadron 'resonances' then emerged: they were essentially different modes of vibration of these strings, rather like different overtones on a violin string.

The first string theory worked only for hadrons belonging to the boson family, but in 1971 a version that applied to fermions was found as well. More intriguing imagery about the subatomic world started to emerge. The reason quarks had never been found outside their host particles is that strings between the quarks prevented their escape. But, as ever, these glimmerings of a new understanding were dogged by technical problems. For a start, string theory for bosons made sense only in a universe of no fewer than 26 spacetime dimensions. Compared to this, Kaluza's one extra dimension seemed positively unadventurous.

There was worse to come. It turned out that when these strings were vibrating in their lowest state, they seemed to generate tachyons – particles that travel faster than light. The appearance of such particles, with all the paradoxes of time travel they evoke, was disastrous.

But by the mid-1970s, hardly anyone cared about the string theory of hadrons in any case. Quantum chromodynamics had emerged as a perfectly decent account of these particles and of the strong force. The problems facing string theory were deemed extremely academic by everyone – everyone, that is, apart from a handful of mavericks who had found hints of something marvellous buried deep in string theory.

Ironically, this clue was once seen as a weakness of string theory. When theorists tried to use it to describe the strong nuclear force, the equations kept throwing up a particle which had no mass and two units of spin. Such a particle had never been seen in any subatomic processes, and this looked bad for a string theory of hadrons. However, in 1974 John Schwarz at the California Institute of Technology and a visiting French theorist called Joel Scherk pointed out that such a particle is demanded elsewhere in physics. The properties it has are just those of the *graviton*, the carrier of the gravitational force. If string theory was no good for describing the strong interaction, perhaps it had something interesting to say about producing a quantum theory of gravity.

Schwarz thought so. He ploughed on with it more or less alone for six years, spending most of his time working with a version of string theory developed by himself, Scherk and some co-workers in 1971. It had two principal virtues – it needed 'only' a ten-dimensional universe in which to work, and it wasn't plagued by time-travelling tachyons.

Finally, in 1979, Schwarz met someone else who had been beavering quietly away on string theory on the other side of the Atlantic.

Michael Green had become captivated by string theory after reading Veneziano's original paper while working on his PhD at Cambridge. By a curious twist of fate, Cambridge was at that time very keen on a rather unconventional approach to particle physics, which was to give Green the technical background he would later find vital.

'My education as a graduate student was extraordinarily poor,' he recalls. 'It was bizarre. There was a period of confusion in particle theory where a whole stream of thought had grown up around the so-called S-matrix theory. This had an inordinate influence on Cambridge, but in the end it proved to be actually a dead end – I don't think there's a single achievement of S-matrix theory which will last. But it had a huge influence in a few places. The result of this was that nobody taught us the run-of-the-mill things everyone else learnt. It was certainly that fact that got me equipped to do rather bizarre things with strings. String theory in the form I was working on is very much an outgrowth of S-matrix theory.'

Green met up with Schwarz over a coffee at the nuclear research centre CERN in Geneva, where they were both hoping to do some theoretical work over the summer. Both knew that they were doing their careers no good at all by pursuing their esoteric fancies in strings. At that time, everyone who was anyone was working on supergravity. Schwarz, more than a decade after getting his PhD, was still a mere research associate at Caltech. Green had only recently landed a steady job at Queen Mary College, London, after almost a decade of short-term appointments. But they both shared a conviction that while everyone else seemed to think supergravity was the way forward, string theory was telling *them* something important about the nature of the universe.

They decided to start their collaboration by following up a suggestion made in 1976 by Joel Scherk and others in one of the very few papers to appear on string theory during those years. This hinted that something interesting might emerge if string theory were combined with the very idea that everyone else was raving about – supersymmetry. Schwarz and Green felt that although supergravity might not be all it was cracked up to be, the power of the supersymmetry idea it incorporated really couldn't be ignored by anyone hoping to build a Theory of Everything.

Their first attempt to forge a link between strings and supersymmetry failed. But in their next attempt they started to make exciting progress. Superstring theory, as the combination was called, seemed to be giving them hints that they would soon be rewarded for their persistence. 'If you're on the correct path, the theory sort of suggests what the answer should be, long before you're anywhere near proving it,' says Green. 'The calculations we were doing at that time were very very complicated – a lot of algebra – and nothing would've motivated me to go through pages and pages of algebra if I didn't think that the end result would be simple.

'We worked together at the same table and in an incredibly intensive fashion – far more intensive than at any other period in my life. I guess being bachelors at the time was a big factor, because it's the sort of subject that once you're working on you can't leave alone. You become totally obsessive.'

But the more work they did, the more amazing the results. Hurdle after hurdle just fell before them; a 'golden road' seemed to be emerging. Green recalls: 'For the first couple of years that we were investigating these theories, we were still just mesmerised by the fact that by studying them in more and more detail, one came up with more and more ways in which these theories were consistent.'

In late 1981, the first major development emerged. Using a boiled-down version of superstring theory, Schwarz and Green found that a simple calculation produced a sensible answer. This might not sound much until one recalls the sheer lack of success any theory that included gravity had had until that date. They all gave nonsensical, infinite answers to even the simplest questions. It seemed that by changing one's view of particles from just pointlike, infinitely small objects to extremely small, yet nonetheless finite *strings*, the new theory could see off the infinities. Admittedly, the result held only for a stripped-down version of superstring theory but it was, they thought, better than nothing.

Unfortunately, no one else seemed to agree. $N=8$ supergravity had yet to die its death and virtually everyone was still concentrating their efforts on it, with big names like Stephen Hawking leading the way. No one seemed to care about superstring theory, despite its apparent resistance to the dread infinity disease.

In 1982, Schwarz and Green made a second breakthrough. For years they had nurtured a hope that *all* superstring theories would be free of infinities. Then presumably experimental predictions would enable just one version to be picked out as the likely theory of everything. In fact, superstring theory was rather cleverer than this. Out of the infinite number of possible superstring theories, only a few turned out to be free of infinities. Mathematics, rather than experiment, seemed to be acting as the sieve, weeding out the hopeless versions of the theory.

In retrospect, this was just as well – it had become clear that experiment might never be able to decide one way or the other on a candidate Theory of Everything. All quantum theories of gravity have a characteristic length on which their effects become noticeable – and it is always extremely small. For superstring theories, that length can be thought of as the length of the superstrings. Their size can be roughly calculated and, sure enough, it turns out to be incredibly small – 10^{-35} metres – smaller even than the curled-up dimension in Kaluza and Klein's theory. The energies needed from a particle accelerator to probe such small distances are literally fantastic. Only once in the whole history of the universe, a mere 10^{-38} seconds after the Big Bang, have such energies existed.

And this raises a philosophical question: how can a theory that demands such awesome conditions for a test of its predictions really command any confidence? This point was later to lead to stinging criticism of superstring theorists by some of the great names in physics.

But in 1982, Schwarz and Green didn't even have so much as a critic. They were still being ignored. 'The lack of interest, I thought, was staggering,' says Green. Yet looking back on those early days, both now view the lack of attention with some relief: at least they could carry on following their noses without worrying what thousands of rivals were doing around the world.

1984 proved to be the decisive year. The hundreds of theorists working on $N=8$ supergravity were finally admitting that their theory was sick. Attempts to raise other versions ran into anomalies and infinities. Meanwhile, Schwarz and Green had decided to attempt a crucial calculation. They knew superstring theory could be free of infinities. Could it be free of anomalies as well? They started the calculation with their hopes high – string theory had so far seen off all the problems that had killed its rivals. Even so, they began cautiously by looking at one particular type of anomaly. As with the infinities, it turned out that the vast majority of superstring theories did suffer from the anomaly. But just one version did not. Once again, the mathematics appeared to be acting as a sieve, selecting out just one, unique theory from countless trillions.

This was exciting, but they had dealt with only one type of anomaly. Could the version they had been left with see off all the other anomalies as well? Their excitement mounting, Schwarz and Green decided to do the big calculation. Green explains: 'For the calculation to work, there had to be a miraculous cancellation between many peculiar-looking numbers so that the answer of adding these numbers together had to be 496.' With Schwarz at the blackboard and Green at a desk, each tried to control his excitement as they did the calculation. Green finally groaned – it seemed the result was different. But Schwarz had another shot, and this time the result was . . . 496.

Somehow, superstring theory had dodged the difficulties that had floored its rivals. It was hard to believe it was not a reflection of some genuine significance of superstrings in the universe. This time, other theorists agreed; they didn't have a whole lot of alternatives to work on anyhow. The lesson for everyone seemed to be: replace your pointlike particles with ten-dimensional, supersymmetric strings.

Some big names also started to take an interest. 'I dropped everything I was doing, including several books I was working on, and started learning everything I could about superstring theory,' recalls Steven Weinberg, Nobel prizewinning co-inventor of the electroweak theory.

One particularly important recruit to the superstring camp was the Princeton physicist Edward Witten. The son of a professor of physics, he originally trained as a historian at Brandeis University, only to realise that theoretical physics was his true calling. He retrained himself and became a theorist held in awe by Nobel prizewinners twice his age. His most impressive quality is his ability to work with equal ease in pure mathematics and theoretical physics. Both mathematicians and physicists regard him as one of their own. 'Ed Witten is a very brilliant man, so it was a splendid moment for string theory when he became involved,' says Murray Gell-Mann.

In 1981, Witten had already carried out some unintentionally useful work for the superstring camp by showing that the rival supergravity with pointlike particles could not be a Theory of Everything. The demonstration was exceptionally – and characteristically – elegant. He showed that supergravity demanded at least eleven dimensions if it were to work properly, but for other reasons could not have more than eleven dimensions. This sounds excellent on the face of it – once again the mathematics has sieved out all but one theory. However, Witten had also shown that only theories operating in *even* numbers of spacetime dimensions could generate the features of fermions that physics demanded. So no supergravity theory could work. There was, however, a get-out clause – the theorists could give up thinking of particles as pointlike objects.

Shortly after Schwarz and Green had shown that superstring theory was free of anomalies, Witten's legendary brilliance made itself felt. Green explains: 'Once we did the proof, it spread by word of mouth to Princeton, where Witten got a very garbled account of it. But he instantly understood and, I'm told, within an hour reconstructed our arguments. He called us up and asked us to send the paper by Federal Express. He got it the next day, and he'd written a paper on it four days later. He was just astonishing.'

Witten is now one of the staunchest supporters of the new developments. 'Superstring theory is a miracle through and through,' he says.

Soon Schwarz and Green found that their work was being moved forward by others almost faster than they could read about them. The new discoveries were every bit as exciting as the two had hoped they would be. A team of four theorists at Princeton in America – which inevitably became known as the Princeton String Quartet – found that it was indeed possible for string theory to account for all four fundamental forces. The trick was to combine the new ten-dimensional superstring theory with seemingly moribund bosonic string theory of the 1960s, with its 'ludicrous' 26 dimensions. This marriage of ideas gave birth to a new mathematical creature for theorists to study: the *heterotic string*.

Rather than being like your average piece of string with two ends flailing about, heterotic strings are actually loops. But the strangest thing about them is the mix of dimensions involved. How can something in 26 dimensions also have ten dimensions? The answer is it can't. Ten of the 26 dimensions are ordinary spacetime dimensions, of which six are curled up, or compactified to invisibility to leave the four that we now observe. The remaining sixteen go to form so-called 'internal dimensions' which enable heterotic strings to account for all the fundamental forces.

Why *six* should be compactified is a mystery. One idea is that in the very early universe, the loops got tangled up with the ten dimensions they inhabited. As a result, the loops prevented some of these dimensions from taking part in the general expansion of the universe that followed the Big Bang. These trapped dimensions thus remained permanently tiny. The other dimensions, however, were blown up to their current size by the expansion process. Mathematicians have found indications that this 'snagging' process might work best when six dimensions end up being trapped by superstring loops. No one knows for sure, however.

The universe in superstrings

With the invention of the heterotic string, we can at last attempt to give a picture of how all the particles and forces in the universe may arise. Down at the unimaginably small level of 10^{-35} metres, looplike heterotic strings are constantly writhing around. Picture a snake wriggling around on a carpet, though, and you miss something vital. Strings do not lie in spacetime – they are *part of it*. A slightly better image is of some vast fabric – spacetime – whose stitches are forever in motion.

But how do strings account for all the particles and forces that we see at work today? According to superstring theory, the particles that make up the universe – both those that carry the fundamental forces and those affected by them – are manifestations of the lowest rate of vibration of heterotic strings. So, on this picture, particles are actually loops of superstring vibrating in their least excited state. The differences in the properties of the myriad particles then stems from the way their characteristic 'charge', be it electrical, weak, strong or gravitational, is smeared around the loop. The theory dictates how this smearing takes place for each type of particle we see.

This, then, is the new picture of elementary particles and the forces of the universe. It may be a Theory of Everything, but it could hardly be more remote from human experience. Can it really be true? Although some of the biggest names in physics have become powerful advocates of the new theory, some equally big names have warned against what they see as the deceptive charms of superstring theory.

The criticisms are essentially twofold. First, the current superstring theory owes its origin essentially not to observation but to abstract mathematics, and the need to get over problems that stopped other putative Theories of Everything dead in their tracks. Thus this latest contender for the coveted title Theory of Everything, say critics, may look very attractive but may be nothing more than a mathematical curiosity with no basis in reality. Even its supporters will admit that no one knows the basic physical principles that lead inevitably to superstring theory. To this extent, everyone agrees that the theory is half baked.

The second criticism strikes even deeper, to the issue of whether or not superstring theory can ever be considered a scientific theory. Most scientists insist that any scientific theory worth the name must make predictions that can be tested. But the tiny scale at which the superstrings exist, and thus the huge energies needed to get to them, seem to put superstrings permanently out of reach of the experimentalist. Many physicists find this lack of contact with the real world disturbing. They claim that superstring theorists are free to do what they like as long as it doesn't lead to mathematical problems.

In the years before his untimely death from cancer in 1989, the brilliant and iconoclastic Nobel prizewinning physicist Richard Feynman became a vocal critic of superstring theorists. 'I don't like that they don't check their ideas. I don't like that for anything that disagrees with an experiment, they cook up an explanation – a fix-up to say "Well, it still might be true" . . . The mathematics is far too difficult for the individuals who are doing it, and they don't draw their conclusions with any rigour. So they just guess.'

This criticism carries all the more force for coming from one of the architects of quantum electrodynamics, the most accurate and successful theory ever devised.

Another powerful critic is Sheldon Glashow, who won a Nobel prize for his work on unification of the weak and electromagnetic forces: 'I'm particularly annoyed with my friends the string theorists because they cannot say anything about the physical world . . . They do not have a theory. They have a complex of ideas which do not evidently form any kind of theory and they cannot even say whether their structure describes the successful accomplishments that have been obtained in the laboratory.'

Glashow has felt moved to poetry by his fear that superstring theory is nothing more than a heap of mathematics being pedalled by some very brilliant theorists:

> Please heed our advice that you too are not smitten
> The book is not finished, the last word is not Witten.

Ed Witten himself is well aware of the criticisms and accepts them. The biggest task, he says, is to find the basic principles that lead to the mathematics. Find them and it may become obvious why six, and not seven or five, dimensions become wrapped up to leave the four we see today. He thinks one of the problems is that superstring theory has come along a bit too early: 'By rights, twentieth-century physicists shouldn't have had the privilege of studying this theory. By rights, string theory shouldn't have been invented until our knowledge of some of the areas that are prerequisite for string theory had developed to the point that it was possible for us to have the right concept of what it was all about.'

The superstringers have responded by saying that they have come up with a prediction of sorts already. Superstring theory *demands* that the graviton exists, and thus that gravity itself exists. Considering that all previous attempts to account for gravity had got precisely nowhere, this is pretty impressive, they say.

Few scientists would reckon this to be a real 'prediction' – after all, we already know gravity exists. Much more impressive is the prediction of something genuinely unexpected. In any case, the superstringers may need some help from their experimentalist colleagues in order to make more progress. In the early days of superstring theory (that is, the mid-1980s) there was a belief that by getting rid of both infinities and anomalies all but one superstring theory – presumably the 'right' one – would be left standing. This is now known to have been a vain hope. There are about three or four different heterotic superstring theories currently under study. At the moment, the underlying principle of physics that will eliminate all but one of these is still unknown. It may be that good, old-fashioned observational data may have to perform its time-honoured role of deciding between the contenders.

There are some hints of 'testability' emerging. Many theories about how strings operate lead to the idea that there may be huge quarklike particles flying around the universe. These particles really are huge by elementary particle standards – about 10^{19} times heavier than a proton or about the same as a bacterium. Because, like quarks, these particles have fractional amounts of electric charge on them, they could be picked out from the background of other particles relatively easily. However, no one knows how many of these quarky particles we would expect to fly through our part of the universe and so there is an understandable reluctance to spend a lot of money searching.

In the meantime, Andrew Liddle and his colleagues at Sussex University have devised an ingenious test of any theory – including superstring theory – which invokes more dimensions to the universe than the four we see. And the results appear to be less than encouraging for aficionados of multidimensional universes. The test is based on peculiar objects first discovered in 1967 by radio astronomers at Cambridge University: neutron stars.

The rapidly spinning remnants of stars that exploded in cataclysmic events called supernovas, neutron stars are about the same mass as the sun, but measure just a few tens of kilometres across. This gives them incredibly high densities – around 10^{15} times greater than steel – and intense gravitational fields. Neutron stars are, therefore, under intense pressure to collapse even further under their own gravitational fields. If they were allowed to – and only subatomic forces prevent it – neutron stars would turn into black holes, objects whose gravity is so strong that not even light can escape their clutches.

Astrophysicists have been able to work out the maximum mass a neutron star can have before it collapses even more to form a black hole. These calculations are, of course, usually worked out on the assumption that there are only four dimensions to our universe. But superstring theory demands that there are in fact many more dimensions – albeit with most of them coiled up so tightly we never see them.

Dr Andrew Liddle and his co-workers looked at what happens if the equations describing a neutron star's interior are worked out for more than the normal four dimensions. To keep things simple, they added on just one more dimension. But the effects were dramatic. The maximum mass that a neutron star can have without collapsing into a black hole is drastically reduced to far below the masses of known neutron stars. So the fact we see *any* neutron stars seems to suggest that there cannot be any extra dimensions to the universe. Does this mean that the extra dimensions of superstring theory really are just a figment of theorists' imaginations? By fiddling with the theory of the interior of a neutron star, it is possible to wriggle out of the problem, but Liddle thinks this smacks of special pleading. He is confident that the existence of neutron stars constitutes a serious problem for any theory with more than four dimensions.

Fortunately for superstring theory, some rather more encouraging news comes from a much more down-to-earth source: particle accelerators. No one expects to get direct confirmation of superstring theory from a man-made machine – the energies involved are far too high. But it is possible to use accelerators to see if there are *trends* in data obtained at relatively low energies towards what one would expect from superstring theory. John Ellis, head of the theory division at CERN, and his co-workers have recently shown that measurements made on CERN's giant LEP machine in Geneva do indeed show a trend supporting the existence of superstrings.

Ellis and his colleagues are also the originators of the most exciting proposal yet for investigating events on the scale of superstrings – a tabletop experiment that might, just might, directly pick up signs of their existence. They appear to have found a way around the problem of needing very high energies to probe events on the smallest scale. The trick consists of using a device built out of so-called superconducting materials, which lose all their electrical resistance when chilled almost to absolute zero, $-273°C$. The device, called a 'superconducting quantum interference device', or SQUID, is based on the work of the Nobel prizewinning physicist Professor Brian Josephson at Cambridge University. At extremely low temperatures, the electrons flowing through superconductors start to behave rather strangely. In particular, their quantum properties start to become very noticeable. These properties are very easy to disturb – just the feeblest amount of magnetism can rock these quantum boats. Ellis and his colleagues believe that the electrons in SQUIDS might actually respond to the writhings of spacetime down at the level of superstrings.

In 1990, they proposed an experiment in which the behaviour of electrons in a SQUID would show signs of being affected by these writhings if closely monitored for half an hour or so. There are, to be sure, plenty of problems with the idea. The theory on which it is based is pretty rough and ready, and could easily be way off the mark. Separating out the incredibly weak effect from more mundane sources of interference would be very difficult. Even so, the mere suggestion of being able to detect the effect of events so removed from everyday experience has proved sufficient for some daring experimentalists to have a shot at it. Terry Clark and colleagues at Sussex University have already set some limits for the size of the effect; whether Ellis's bold proposal will pay off has, however, yet to be seen.

It may, of course, turn out that superstring theory is what many of the old guard in physics believe it to be: an intensely mathematical chimera. At the beginning of this brief study of the Outstanding Mystery of the Theory of Everything, the problems of treating gravity as 'just another quantum field' were cursorily described. These problems could, however, be genuine and insurmountable. But in the face of the appalling mathematical difficulties that beset the building of a Theory of Everything, theoretical physicists remain a sanguine tribe. There *are* some powerfully attractive mathematical features of superstring theory, suggesting that it may truly be a fundamental part of the grand scheme of things. There *are* glimmerings of support for superstring theory from particle accelerators.

But even if it all works out, will we really have a Theory of Everything? To those outside theoretical physics, the answer is clear: absolutely not. How do we explain the genius of Mozart? The pleasure we get from watching the sun set? The reason we are here? This is, however, to misinterpret what physicists mean by *everything*. All the matter in the universe and all the forces that act on it is 'everything' to a physicist. From these, everything – including the thoughts inside our heads – ultimately flows, though the route may be tortuous indeed.

Much harder to avoid are the criticisms levelled at Theories of Everything by physicists themselves, such as John Barrow of Sussex University. He warns that theorists have ultimately spent their time trying to understand the physics of a situation that simply no longer exists. The four fundamental forces *may* once all have been a single 'superforce' in the very early universe, but they are patently *not* today. Predicting how we ended up in our particular situation in this part of the universe – as one might reasonably expect a Theory of Everything to do – would involve knowing how these forces interacted with matter during the intervening fifteen billion years. And that could involve a substantial amount of randomness, whose effects would be impossible ever to work out.

Barrow's criticisms of Theories of Everything are not based merely on assumptions of how clever we can be – they stem from fundamental properties of mathematics, the very tool the theorists use. Chaos theory teaches us that even very simple processes can lead to utterly unknowable outcomes. But more fundamental work on the foundations of mathematics has also shown that there are some questions whose truth can never be established using the laws of logic. In the face of such insuperable barriers to knowledge, claims of the discovery of a Theory of Everything have a hollow ring.

Michael Green, one of the pioneers of superstring theory, shares some of the sentiments of those who would have the theory viewed in a more subdued light: 'Saying it's a Theory of Everything is merely saying it seems as though it might answer the questions that we now think are important in particle physics.'

Ed Witten, arguably the most brilliant of the current generation of theorists, sees no reason for circumlocution, however: 'Super-strings are either *the* theory of nature, or they're an incredible step forward toward the theory of nature.'

It is somewhat ironic that gravity, that most familiar of all the fundamental forces of nature, should cause theorists so much trouble in their search for the Theory of Everything. Yet its familiarity masks the fact that gravity forms a bridge between the two frontiers of scientific knowledge: superstrings and cosmology. At the one extreme, quantum gravity dictates the size of the very stitches in the fabric of spacetime. At the other, gravity dictates how the vast, universal sheet of spacetime – our universe – behaves. In the next, and final, chapter, we will uncover the power of gravity to both shape our universe – and to end it.

6   The Universal Mysteries

I N 1835, THE FRENCH PHILOSOPHER Auguste Comte declared that no one need try to understand the universe. 'The field of positive philosophy', he opined, 'lies wholly within the limits of our solar system, the study of the universe being inaccessible in any positive sense.'

Fortunately, scientists in the nineteenth century took not a blind bit of notice of what Comte had to say and, indeed, proceeded to demonstrate the fatuity of his views. The result was the development of the most imposing of all the fields of modern science: cosmology.

The aims of cosmology are almost risibly ambitious: to understand the origin, nature and ultimate fate of the universe. Even physicists, who have something of a reputation for arrogance among scientists working in other fields, have been known to look askance at cosmology. Lev Landau, a brilliant Soviet theoretical physicist, once described cosmologists as being 'often wrong, but never in doubt'. Certainly current theories about the nature of the universe are still in a state of flux. The great difficulty of acquiring data about the universe as a whole has made clear-cut predictions – the test of a real scientific theory – hard to come by. Even Einstein managed to make a botch of one. But considering the scale of the task before us, the progress that has been made is truly astonishing.

The apple and the universe

It is gravity that controls the universe. Although by far the feeblest of the four fundamental forces at work within the universe, gravity's intimate relationship with matter leads to its complete domination of the physics of the very largest objects: galaxies, clusters of galaxies and the entire cosmos. Understanding gravity, therefore, is the key to unlocking the mysteries of the universe.

Modern cosmology is considered to have begun with Sir Isaac Newton and his discovery of the law of gravity in the 1660s. By 1687 Newton had found that his law applied far beyond the apple trees of his mother's Lincolnshire garden. He had extended it out as far as Saturn, then the most distant known planet in the solar system.

A letter from Newton, written in 1692, shows that by then he had begun to push his theory further still – out beyond Saturn to encompass the entire universe. This intellectual bound was made almost 150 years before Comte made his defeatist declaration. Newton began modestly enough by considering what would happen if the universe were very large but not infinitely so. In doing so, the first of the many conceptual problems that bedevil cosmology raises its head: how can a universe which, by definition, contains everything be *finite*, and thus have something 'beyond' it? But Newton found there were mathematical problems, too. Gravity, according to his own law, attracts each and every particle. Thus eventually every particle should end up in some huge ball of incandescent matter at the 'centre' of the universe. In a universe that has existed forever, such a situation must have already come to pass – but it patently hasn't.

Newton got round both the semantic and scientific problems in one go by thinking of the universe as truly infinite. Then, said Newton, matter 'could never convene into one mass; but some of it would convene into one mass, and some into another, so as to make an infinite number of great masses scattered at great distances from one another throughout all that infinite space. And thus might the Sun and fixed stars be formed.'

In hindsight, this vision of a static, infinite universe dotted with bright objects looks like a brilliant prediction of the existence of galaxies, more than 200 years before they were clearly identified. But in blithely talking of infinities, Newton had shown scant regard for the hazards they hold for all physicists. Infinities are dangerous mathematical creatures, alive with paradox. And so it proves with Newton's impressive conception of a gravity-filled universe. The problem boils down to this: think of the universe as filled with stars all of the same mass, evenly dotted about space. Now, if one tots up the gravitational force on any particular star in the universe, one gets a pull from one particle on one side and an equal and opposite pull from one on another, and so on and on forever (this is an infinite universe, remember). So, to work out the net force on a star, taking into account all these contributions, one ends up with an addition to carry out. This sum boils down to the form

Sum = 1 − 1 + 1 − 1 + 1 − 1 + 1 − 1 + etc . . . forever.

Let us work out the value of this sum. To make things easy, we could package it up thus:

Sum = (1 − 1) + (1 − 1) + (1 − 1) + (1 − 1) + . . . forever.

This is exactly the same as the original; all we have added are brackets, to make things clearer. We find that all the brackets add up to zero, so Sum = 0, and the net force we get will also be zero.

But wait − we could package up the sum in another way:

Sum = 1 + (−1 + 1) + (−1 + 1) + (−1 + 1) + . . . forever.

This is also identical to the original sum, and again all the brackets add up to zero. But this time we still have a 1 left over at the beginning, so the value of the sum is 1, and not zero − a rather different result! This illustrates one of the dangers of dealing with infinite quantities − one never knows exactly how to package them up, as one never knows when one has got hold of them all.

And so, in trying to understand the behaviour of the whole universe, Newton's theory of gravity finally dies − as so many much later ones have died − from the infinity disease.

Einstein and the equations of the universe

Not until 1915 and the advent of an entirely new concept of gravity did an answer emerge. Einstein's general theory of relativity (GR) showed that gravity is not some sort of mysterious 'elastic', binding each particle in the universe to every other. According to GR, gravity is a manifestation of the curvature of space and time. Thus the sun keeps the planets in orbit about her by curving the spacetime in which they exist, rather like a ball sitting in some vast rubber sheet.

As we learned in the last chapter, in 1915 Einstein had the evidence he sought for the correctness of this new view of gravity: the equations of GR neatly explained some peculiar behaviour in the orbit of Mercury which had defied explanation by Newton's theory. Less than two years later Einstein was ready to turn his theory to the problem of the origin and fate of the universe.

Clearly, GR had to be free of the problems of the infinite that crippled Newton's theory. It also had to agree with what was known about the universe. And in 1917, the consensus was as it was in Newton's day – that the universe was *static*.

Clues of how to get GR to meet these two demands came from Newton's old law. It turns out that both the infinity problem and the need to give a static universe can be solved by assuming that there is more to gravity than just the famous 'inverse square' law of Newton, by which the strength of gravity falls with increasing distance. If one assumes that there is a secondary effect of gravity, and one that grows in importance the further one ventures into space, it is possible to keep all the successes of Newton's theory while also curing its 'cosmic' problems. All one must do to Newton's equation is to add a little modification.

Einstein found that the equations of GR did not, however, include this extra, long-range effect. But then, these equations had not been applied to the universe as a whole. Once they were, was it not reasonable that they might have to be modified?

Einstein therefore gave arguments for why one should insert into GR exactly the same type of modification as that needed to get Newton's theory to work. It is known as the *lambda term*, and the strength of the effect is measured by the so-called *cosmological constant*.

The equations of GR are indeed free of the 'Newtonian' infinities. This is essentially because it includes the key new concept of gravity causing space and time to curve, an idea that, understandably enough, never occurred to Newton. The equations of GR also gave a universe that is static.

But Einstein soon realised that there was something wrong with what he had done and grew uneasy about the modifications he had made to the 'raw' GR equations which had, after all, given him his greatest successes.

Einstein was right to feel uneasy about the lambda term; it would rob him of what would arguably have been his greatest triumph. For in 1912, even as he was groping towards the original equations of GR, an astronomer at an observatory in Arizona was taking measurements that would lead to a revolution in our view of the universe – a revolution that Einstein could have predicted, had he held faith with GR unsullied by this lambda term.

The discovery of the universal expansion

The story of perhaps the most astounding scientific discovery ever made begins around 1860. It was then that the German scientists Robert Bunsen (of burner fame) and Gustav Kirchhoff discovered how to blow a hole in Comte's philosophy. They found that each chemical element had a characteristic 'fingerprint' – a spectrum – that emerged when the element was heated and the resulting light passed through a glass prism.

Soon, astronomers were turning their 'spectroscopes' on to the distant stars. In 1863, a British astronomer named William Huggins discovered that the spectrum of the stars bore similarities to those generated by incandescent chemical elements on Earth. He had shown that the spectroscope could do what Comte thought impossible: enable us to know the nature of objects which lie forever beyond our reach. By 1868, spectroscopy had uncovered an element on the sun that was unknown on Earth – it was named *helium*, and was not discovered on our planet until 1895.

But the most spectacular discovery was yet to come. In 1912, Vesto Slipher, an astronomer at the Lowell Observatory in Arizona, turned the observatory's telescope on a cloud of gas and dust in the constellation of Andromeda. His plan was to investigate the chemical composition of this spiral-shaped cloud. After exposing a photographic plate to the feeble light of the cloud, Slipher went back to the lab to find the telltale fingerprint of the chemical elements. Sure enough, many of the spectral features of Earthbound chemical elements appeared in the nebula's spectrum, showing that the Andromeda nebula was built out of the same hydrogen, helium and other elements found in our solar system.

But there was something else. Although all the features making up the fingerprint were in their correct *relative* positions, the *entire* fingerprint had somehow been moved. All its lines had been shifted towards the blue higher-frequency end of the rainbow. Slipher knew the explanation: he was seeing the Doppler Effect.

Most people are familiar with the Doppler Effect by sound rather than sight. When a noisy car approaches, the pitch of the sound increases – that is, the basic engine noise is shifted towards higher frequency. Similarly, as the car goes away from us, the pitch drops.

For a source of light rays, the details are slightly different but the result the same: a Doppler shift boosts frequencies when the source is coming towards us. Slipher's discovery of a wholesale shift towards the blue (higher-frequency) end of the spectrum of the Andromeda nebula thus had a simple interpretation. The whole nebula is moving towards us. But Slipher got something of a shock when he calculated just *how* fast the nebula was approaching: 720,000 kilometres an hour. This is an incredible speed; similar measurements made on individual stars usually reveal far lower speeds.

Intrigued, Slipher set about building up a catalogue of speeds for other nebulae. By 1914, he had obtained and analysed the spectra from thirteen of them. It turned out that most had red shifts in their spectrum – in other words, they were racing away from us. And again the speeds were astonishing: more than a million kilometres per hour – the highest speeds then measured in astronomy.

Ten years later, the truly cosmic significance of Slipher's discovery started to emerge. In 1924 the American astronomer Edwin Hubble, working with the 100-inch telescope at Mount Wilson, California, discovered that the nebulae were not part of our galaxy at all, but belonged to separate galaxies lying far beyond our own. This implied that whatever Slipher's measurements meant, their relevance was not confined only to our galaxy. They had implications for the entire universe.

The crucial breakthrough came in 1929. Hubble announced that he had found a law between the size of the spectral shifts and the distance of the galaxies. He had discovered that the further away a galaxy was, the faster it raced away: this is Hubble's Law; see top diagram, opposite. The ratio of the speed of recession of a particular galaxy to its distance from us is a quantity that has since taken on central importance in cosmology. Some astronomers have devoted their entire careers to getting an accurate value for this figure. It is known as *Hubble's constant*, H.

Hubble's discovery raises a question: why are all the galaxies racing away from us? It turns out that, in fact, any observers sitting in those distant galaxies would also get the impression that everything is racing away from them. The best way to visualise the situation is to think of a balloon with small coins stuck on to it; see lower diagram, opposite. Imagine the galaxies as coins stuck on the skin of a balloon representing the space between them. As the balloon is inflated, the coins all start to move away from one another. Seen from any one coin, the expansion of the balloon seems to be taking place away from that coin – but in fact all the coins are moving

Hubble's Law: how the speed of recession of distant galaxies increases with distance

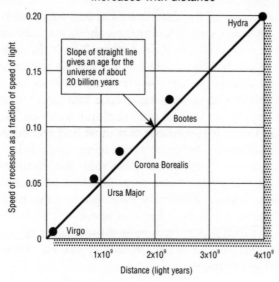

Slope of straight line gives an age for the universe of about 20 billion years

- Hydra
- Bootes
- Corona Borealis
- Ursa Major
- Virgo

Speed of recession as a fraction of speed of light

Distance (light years)

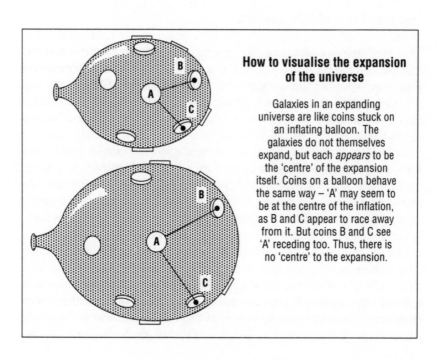

How to visualise the expansion of the universe

Galaxies in an expanding universe are like coins stuck on an inflating balloon. The galaxies do not themselves expand, but each *appears* to be the 'centre' of the expansion itself. Coins on a balloon behave the same way – 'A' may seem to be at the centre of the inflation, as B and C appear to race away from it. But coins B and C see 'A' receding too. Thus, there is no 'centre' to the expansion.

away from all the others. There is no unique 'centre' to the expansion of the balloon's surface.

Some galaxies, such as the Andromeda spiral, are relatively close to us. As a result, gravitational forces between the two galaxies leads each toward the other. The result is a blue shift in the spectrum of the galaxy. But the key point about Hubble's 1929 announcement is very simple: the *distant* galaxies all move away from us. Thus we are led ineluctably to the astounding conclusion that *the universe as a whole is expanding*.

Clearly, one of the central assumptions Einstein had made in formulating his equations of the universe, based on GR, was wrong. The universe was not static. There was, therefore, no need to introduce the lambda term. The original equations of GR had been sufficient all along. If Einstein had used them in their 'pure' state, unsullied by the lambda term, he could have made arguably the most impressive prediction in the history of science: that the universe expands. Small wonder that in later years Einstein referred to his use of the lambda term as 'the biggest blunder of my life'.

The measure of the universe

The now defunct idea that the universe is static had its origins in Aristotle's recognition that anything else led to awkward questions about the *origin* of the universe. With the discovery of Hubble's law, the question of the birth of the universe became of central importance. In particular, the value of Hubble's constant, H, enables one to calculate the age of the universe.

Following Hubble's discovery, astronomers set about getting an accurate value for H. It soon turned into one of the most contentious endeavours in the whole of science. Two sets of data are needed to work out H, and thus the age of the universe. The first is the velocities of distant galaxies. This is relatively straightforward – spectroscopic red shifts have long been automatically and accurately measured.

The other necessary information is the corresponding distances of these galaxies from us. And this is where the problems start. The essence of the problem can be simply explained. Suppose you have a 100-watt bulb shining in a lamp. Put the lamp a few hundred metres away, and the bulb looks fainter, even though it is still consuming 100 watts of power. There is a simple rule that gives the ratio of the perceived, *apparent* brightness of an object to its real, intrinsic brightness; this is known as the 'inverse square' law. It states that if you put a source of light, say, three metres away and measure its apparent brightness, then move the source back to 30 metres and measure it again, the apparent brightness will have decreased by the square of the ratio of the distances. Thus, in this case, the reduction in apparent brightness will be 30/3 squared. In other words, apparent brightness will be reduced by a factor of 100. Move the source back to 60 metres away, and the apparent brightness will have been reduced by 20^2, i.e. by a factor of 400, and so on.

The same basic idea can be applied to entire galaxies. A galaxy 500 million light years away should – all other things being equal – be 100 times fainter than a similar one just 50 million light years away. Thus, by measuring the apparent brightness of a galaxy whose intrinsic brightness is known and applying the inverse square law, its distance can be worked out.

Although simple in principle, this technique is notoriously difficult to apply in practice. For example, how can one ensure that 'all other things' are equal? In particular, how does one choose a standard galaxy whose intrinsic brightness is reliably known? Is there dust or other matter between us and the galaxy that might dim it more than its distance would predict? Does the intrinsic brightness of a galaxy change with time and, if so, how?

Not surprisingly, the first attempts to determine the size of Hubble's constant, and thus the age of the universe, turned out to be hopelessly wrong. The first measurements of distance are now thought to have been too low by a factor of ten – in other words, they indicated that the universe was expanding ten times more rapidly than it really is. This in turn gave a value for the age of the universe that was too low, so low in fact that it turned out to be shorter than the age of the Earth as estimated by radioactive techniques applied to rocks.

The bizarre idea of a universe younger than the Earth led to the emergence of the controversial 'steady-state' theory put forward by Thomas Gold, Herman Bondi and (in a slightly different form) Fred Hoyle at Cambridge University in the late 1940s. They showed that the existence of Hubble's law does not, in fact, demand that there must have been a beginning to the expansion. By thus severing the link between Hubble's constant and the age of the universe, the theory allowed the universe to expand forever, and so got around the age problem. The theory got its name from the fact that it was based on the idea that the universe's current state is the state it has always been in.

The steady-state theory is, in mathematical terms, wonderfully elegant and leads to ideas that are at centre stage in cosmology today. But, sadly, by the end of the 1950s, new determinations of Hubble's constant had revealed the flaws in the earlier observations. The result was a figure for the age of the universe that was greater than that of the Earth after all.

Yet the debate about the actual value of H refuses to die. Current estimates of the date of the creation lie somewhere between 15 and 20 billion years ago. Although comfortably greater – by about five billion years – than the age of the Earth, there is still concern that some estimates of H indicate that the universe is younger than some clusters of stars within it.

There is, however, little doubt now that the steady-state theory is no longer tenable. The universe really did have an origin some time in the distant, but not infinitely distant, past. And the proof of it came in the work of two engineers working for a telephone company.

The discovery was made during research in an area seemingly far removed from the esoteric realms of cosmology. In 1961, a telecom-munications engineer at the Bell Telephone Laboratories in Holmdel, New Jersey, was checking out a giant horn-shaped radio aerial designed to pick up the very short-wavelength radio waves emitted by communications satellites. These signals are pretty faint and so engineers are always battling to minimise the amount of background 'noise' that threatens to swamp them.

The Bell engineer was having trouble with the new aerial. When it was pointed at the sky, it seemed to pick up a surprisingly large amount of this noise. Some calculations suggested that the source was probably inside the antenna, so the engineer wrote up a report on the new antenna and thought no more about it.

Then, in the late spring of 1964, along came two more engineers, named Arno Penzias and Robert Wilson, to work with the aerial. They too found that it seemed to be picking up rather a lot of background noise. But this time, they tried to track down its source by monitoring the way it varied over time. They reasoned that if the noise reached a peak every 24 hours, perhaps there was some object in the sky that might be responsible. Curiously, however, the noise stayed constant night and day, regardless of the part of the sky at which the antenna was pointed.

As the noise didn't seriously affect their work, Penzias and Wilson simply noted the discovery of the peculiar effect and carried on with the task of calibrating the antenna. Every now and again, though, they would remeasure the level of noise. It was always the same.

Somewhere, it seemed, there was a feeble extra source of heat, a mere three degrees or so above absolute zero, $-273°C$, feeding energy into the antenna. Perhaps the source of heat was inside the antenna itself. The scientists noticed that birds fluttered around the antenna sometimes; were their droppings having some effect? They opened up the throat of the antenna and cleaned its surfaces. The noise did indeed decline, but only by about ten per cent or so; the bulk of it remained. Having ruled out a host of Earthbound explanations and apparently astronomical ones as well, Penzias recalls, 'We frankly did not know what to do with our result.'

The first hints that they had stumbled across something of literally cosmic importance came during a telephone call Penzias made to a colleague. It emerged that Professor Robert Dicke and colleagues down the road at Princeton University had been investigating events taking place around the time of the birth of the universe and had reached a very interesting conclusion. They believed it would still be possible to pick up the heat left over from the cataclysmic explosion – the Big Bang – which began the expansion of the universe. Because of the Doppler red-shift effect, the heat radiation would be shifted to longer wavelengths, into the short radio or 'microwave' band. The Princeton team had even been able to calculate the corresponding temperature of the microwave remnant of the heat: about ten degrees above absolute zero.

The Princeton team had begun to set up their own experiment to detect this heat when they heard the news from Bell Laboratories. Penzias and Wilson had, it seemed, found the heat of the Big Bang. Suddenly, all the pieces of the puzzle fell into place. The reason the strength of the noise never seemed to vary was that the whole of space was filled with its source, the dying embers of the Big Bang. Quickly the two teams got together at Bell Laboratories and decided to publish a joint paper, with Dicke's team making the prediction and Penzias and Wilson describing their discovery. For their discovery, Penzias and Wilson would win the 1978 Nobel prize for physics.

In fact, decades earlier, a number of other scientists had come within an ace of making the same discovery. In 1940, Andrew McKellar of the Dominion Astrophysical Observatory in British Columbia had argued that it might be possible to explain certain features of the spectra from distant stars on the assumption that the whole of interstellar space was at a temperature of 2.7°C above absolute zero. McKellar made a prediction, based on this idea, which was quickly confirmed. But unfortunately there were a lot of other possible explanations as well, so despite the successful prediction, McKellar's theory was written off by one Nobel prizewinner at the time as having 'a very restricted meaning'.

A little later, in the late 1940s, three theorists at Johns Hopkins Applied Physics Laboratory in Maryland also showed that if the expansion of the universe had started in an explosion, some residual heat should still persist. They calculated that the current temperature should be no higher than five degrees above absolute zero – just as Penzias and Wilson found it to be years later. Yet the Johns Hopkins team seem not to have realised that the heat would have had special characteristics that would have enabled it to be picked out from all the rest of the background energy filling the universe.

In 1964, two Soviet physicists not only predicted the existence of the heat radiation but even said it should be looked for using the very antenna Penzias and Wilson were working with. Incredibly, they fluffed their chance of making one of the greatest discoveries of all time by misreading the original Bell engineer's report and concluding that there was something wrong with the basic theory of events in the early universe.

Professor Robert Dicke thinks that, in retrospect, the heat of the Big Bang could have been detected as early as 1946, had there been enough interest. But in those days cosmology was the domain of just a handful of the most adventurous theorists far from the mainstream of physics. As Cornell University astrophysicist Martin Harwit puts it: 'The effort required would surely have been horrendous, and perhaps no one could have become sufficiently motivated when there were so many other, simpler observations to be made in astronomy.'

The secrets of the cosmic embers

The discovery of the 'microwave background' by Penzias and Wilson was a turning point in the science of cosmology. Its mere existence convinced almost everyone that questions about the origin of the universe could no longer be evaded. The microwave background also provides cosmologists with one of the very few hard pieces of data they have about events in the early universe. Not surprisingly, since 1965, enormous efforts have been made to measure its properties as accurately as possible, and then to bleed them of as much information as possible.

For years, definitive proof that the microwave background had the characteristics expected of the embers of the Big Bang was lacking, because astronomers were unable to get high enough above the Earth's atmosphere to avoid its confusing effects. But by 1989, a satellite was put into space specifically to measure the microwave background. Called the Cosmic Background Explorer (COBE), its findings are in simply astonishing agreement with theory. If the universe were once much hotter and denser than today, one would expect that there would have once been a time when the whole universe was one giant ball of seething radiation and matter, all at the same temperature. Such conditions are a sure recipe for producing black body radiation – and such radiation always shows a distinctive relationship between intensity and wavelength. COBE found that the cosmic radiation follows a curve whose agreement with the black body law can hardly be better (see diagram on next page).

COBE was also able to measure the temperature of the universe with far greater accuracy than ever before. It found a figure of 2.736°C above absolute zero, measured to an accuracy of about one-fiftieth of a degree. Taking this figure and an estimate of the density of matter in the universe today, theorists have been able to predict just what the universe should now be made of. Questions about the ultimate source of all the chemical elements first emerged after Lord Rutherford and his contemporaries had uncovered the constituents of atoms. For years, theorists were stuck for an explanation of the origin of the elements. It seemed that enormous temperatures – far higher even than the 60 million degrees of the centre of the sun – were needed. But by the 1940s, a few scientists were thinking that the answer might lie in a far hotter place – the early universe. In 1946 the Russian-born American physicist George Gamow had started working on this new possibility with a graduate student named Ralph Alpher. They began with the assumption that the universe was initially comprised of just neutrons and photons. This was a clever choice: left to themselves, neutrons disintegrate after twelve minutes

The proof that the universe began in a Big Bang

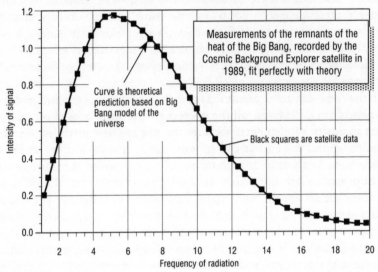

Measurements of the remnants of the heat of the Big Bang, recorded by the Cosmic Background Explorer satellite in 1989, fit perfectly with theory

Curve is theoretical prediction based on Big Bang model of the universe

Black squares are satellite data

Intensity of signal

Frequency of radiation

or so to produce a proton, an electron and an antineutrino. The protons and neutrons can then come together to produce the nucleus of the hydrogenlike material deuterium.

These so-called 'deuterons', having like charges, would normally repel one another. But the intense heat of the early universe smashes them together with such violence that they fuse, producing helium as the main end product. Gamow and Alpher calculated that the end result of these 'nuclear fusion' processes would be that virtually all the matter in the universe should be in the form of helium. Unfortunately, the most abundant element in the universe is hydrogen, which makes up about 75 per cent of the mass of the universe; helium accounts for only about 25 per cent. What had gone wrong? In 1950, a Japanese astrophysicist reworked the original theory, including some effects that had been ignored by Gamow. This brought the amount of helium down to about 40 per cent of the universe by mass – closer to the observed value, but still not good enough.

What was needed was a really detailed calculation that took into account all the particle interactions that would have taken place in the primordial inferno. This job required considerable computing power and was not carried out until the mid-1960s. By the end of

1966, Jim Peebles of Princeton had made the new calculations. The helium abundance that dropped out was 27 per cent – essentially perfect agreement with observations. Other researchers, notably the British astronomer Fred Hoyle and the Americans William Fowler and Robert Wagoner, quickly confirmed the results.

Gamow's original dream, however, was that the Big Bang might account for the creation of all the chemical elements, from hydrogen all the way up to uranium. This proved impossible; the universe expands too quickly for all the necessary reactions to take place. The vast majority of elements, such as the oxygen and nitrogen you are breathing and the carbon in the wood, plastic and steel around you, were born in another cosmic furnace: a supernova explosion.

Supernovas are the death throes of giant stars. After billions of years of steady fusion 'burning' of hydrogen to produce helium, these stars start burning helium as well. The result is the formation of ever heavier, more complex chemical elements. But this process cannot continue forever. When elements as heavy as iron have been produced in the star's core, the whole star runs into trouble. The core can no longer support fusion reactions and starts to collapse, forcing up the core temperature. The outer layers, meanwhile, 'realise' the core has collapsed, and start to collapse themselves. They soon catch up with the core and bounce off it, creating a shock wave that shoots upward through the star, boosted by the extra heat energy from the incredibly hot core.

About an hour after the core collapses, the shock wave bursts out the surface of the star and tears it apart in a truly cataclysmic explosion. The supernova (code-named SN1987) observed by astronomers in 1987 briefly had an energy output equal to all the other stars in the visible universe.

But amid all the destruction, the supernova gives birth to the heavy elements that the original Big Bang could not itself create. These elements are formed by the passage of the searing shock wave through the star; the subsequent explosion throwing the products out into the surrounding void. Thus everything you see around you now is made up of chemical elements that were created in the death throes of some huge star billions of years ago. We are all, to coin a hippy phrase, truly 'children of the stars'.

The particle connection

The creation of the right amounts of the light chemical elements is regarded as a triumph for the Big Bang theory. But the calculations are based on the assumption that the present temperature of the universe is about three degrees above absolute zero. But is it possible to work out the temperature of the universe? The answer is yes. In so doing, we will reach a solution to another deep mystery of the universe. Oddly enough, the theory to do all this comes from research into physics taking place on the subatomic scale.

The mystery is simply stated: why is there so little antimatter in today's universe? Paul Dirac, the British physicist who predicted the existence of antimatter in the 1920s, showed that all elementary particles have corresponding antiparticles. Yet we know there is relatively little antimatter in the universe today because whenever matter meets antimatter they destroy each other in a burst of radiation. Measurements of the background of such radiation has enabled astronomers to put limits on the amount of antimatter in our galaxy: it cannot exceed more than about one part in a thousand of all the matter within it. Presumably roughly equal amounts of matter and antimatter were formed in the original Big Bang. So what happened to it all?

Enter the grand unified theories. As we saw in the last chapter, these predict the existence of a new, extremely heavy particle, dubbed the X-particle, which flits between the quarks that make up particles like the proton and the neutron. It is the X-particles that are thought to cause protons – and thus all atoms – eventually to disintegrate (though proof of this is still lacking).

But X-particles can themselves disintegrate into cascades of matter and antimatter particles – and so can their antimatter counterparts, the anti-X particles. And here is the trick for solving the mystery of the matter/antimatter imbalance. Although both types of X-particle would have been born in equal numbers in the early universe, the two species do not, theorists believe, disintegrate at exactly the same rate. As a result, their matter and antimatter offspring could not cancel each other out in the early universe – and the result was the preponderance of matter we now see today.

But GUTs can go further than this. Although not all the offspring particles were 'paired off', the vast majority were, leading to huge quantities of intense gamma radiation. Red-shifted into longer-wavelength microwaves, this radiation was, according to GUTs, detected by Penzias and Wilson in 1965. Some theorists have attempted to predict just how high the remnant temperature of the pairing-off process should be. To be sure, there is a lot of judicious juggling of figures involved, but it is at least possible to arrive at a figure of around three degrees absolute – just as observed. Although the possible range of temperatures predicted by GUTs is huge, the fact that a theory of subatomic particles can get anywhere near predicting some feature of the entire universe is impressive.

Too good to be true?

The microwave background has one more characteristic of cosmic significance: its astonishing *uniformity*. Penzias and Wilson found that the noise their antenna was picking up seemed to be of the same intensity regardless of the part of the sky at which the antenna was pointing. In other words, the temperature of the universe seemed to be the same in all directions. More than twenty years of further observations of the background have failed to find any unexpected variations in the cosmic background. The latest measurements show that it varies by only part in 50,000 over large areas of the sky.

This incredible degree of smoothness is beginning to worry many astrophysicists. It means that 300,000 years after the Big Bang, when matter and radiation 'decoupled', and were no longer in equilibrium, the matter in the universe was spread throughout the cosmos very smoothly indeed. But how could the universe attain such smoothness so soon after its cataclysmic creation? Was it born smooth or did it have smoothness thrust upon it?

Those most worried about the evenness of the background radiation are astrophysicists grappling with one of the major mysteries of astronomy: the origin of the galaxies. These huge collections of dust, gas and stars are spectacular evidence that the universe today is far from smooth.

Back in 1902, the distinguished British astrophysicist and populariser of science Sir James Jeans found a way of generating such lumpiness from an initial 'seed' of a certain mass. If the seed exceeds this mass – the Jeans mass – it will collapse under its own gravity, overcoming its own resisting pressure. In 1946, a Soviet theorist modified Jeans' work to take into account the effect of the expansion of the universe, which tends to prevent any collapse. This enabled theorists to estimate how lumpy the universe must have been 300,000 years ago to give the lumpiness we now see. Disturbingly, it turns out to have been considerably greater than the smoothness of microwave background will allow.

Thus, simply expecting small amounts of lumpiness in the early universe to build up because of gravity into big, galaxy-sized lumps will not work. But trying to find a way of producing galaxies that does work is proving extremely difficult.

Again, work carried out in the early 1980s suggests that GUTs may be the unlikely source of an answer to the mystery of galaxy formation. In the very early history of the Universe, tiny quantum fluctuations were generated which were then boosted up into much larger-density fluctuations. The simplest versions of GUTs predict a range of different types of fluctuation that more or less matches those needed to produce galaxies. The problem is that it also gives a range of densities for the fluctuations that runs foul of the observed smoothness of the microwave background. But then, we know from the proton-decay experiments that this simplest GUT is wrong. Perhaps supersymmetric GUTs will overcome this problem. Then it might be possible to predict the existence of galaxies using just pencil, paper and particle physics.

Although the enigma of galaxy formation remains, there is now little doubt that the universe did indeed begin in a Big Bang about fifteen billion years ago. So far, however, we have been concentrating on events that took place *after* the Big Bang. It is now time to address the first of the Outstanding Mysteries of cosmology: what *triggered* the Big Bang?

OUTSTANDING MYSTERIES
How did the universe begin?

Ever since Hubble made his pioneering observations during the 1920s, we have known that the universe is expanding, each galaxy racing away from every other. By measuring the rate at which the galaxies are receding, astronomers have estimated that the whole process began about fifteen to twenty billion years ago. This, then, is the date of the Big Bang – the colossal explosion in which all matter was created.

But what started the Big Bang?

For decades, this seemed a question beyond the reach even of Einstein's theory of gravity, general relativity. The theory can take us back almost to the very beginning of the universe but then it breaks down just at the most interesting point: the creation itself.

On the question of what happened at the beginning itself, Einstein's theory is worse than silent: it talks gibberish. This was proved in the 1960s by two British theorists: Roger Penrose, now at Oxford, and Stephen Hawking, now at Cambridge and best known for his work on black holes. They showed that unless weird processes unknown to Einstein were at work at the very beginning, his theory ends up being consumed in a *singularity*. This is a particularly nasty example of the infinity disease that plagues so many theories in physics. For cosmology, its appearance is generally taken to mean that – according to GR – the universe began in a single point of zero size into which all the matter in the universe was crammed. Temperatures, density, strength of gravity – all soared off to infinity at the creation, according to GR.

However, we have met infinities before. We have learnt that they often mean that something crucial has been missed out of whatever theory has gone haywire. So what could be missing from general relativity? The theory seems to have trouble dealing with the early stages of the universe. Working the present expansion back in time, the universe must have been far smaller then than it is now. We know that it is when trying to account for the time the universe was supposedly point-sized that the theory goes haywire. And this is a big clue to what is missing. Theorists have long known that peculiar things can happen when one deals with physics on the smallest scales. This is the domain of the quantum and when talking about events on tiny scales, quantum theory must be used. But general relativity is built on 'classical', pre-quantum theory notions of space and time. Small wonder, then, that Einstein's theory breaks down when the universe was tiny and close to its birth.

Does quantum theory provide a way out of the infinities? To answer this question, a full quantum theory of gravity is needed. Despite growing optimism over the superstring theory outlined in the previous chapter, it is not clear that the theory really is the holy grail everyone is seeking. But the consensus among theorists is (as ever) an optimistic one: that quantum gravity will indeed be free of the singularity problem.

The reason, broadly speaking, is that Heisenberg's famous uncertainty principle should come to the rescue. Applied to the universe, it implies that it is impossible simultaneously to measure the position and rate of expansion of the universe to any desired accuracy. In other words, the uncertainty principle 'smears out' the pointlike precision of the singularity into an incredibly small but nonetheless finite-sized region. Calculations about events in this incredibly small smudge are difficult but no longer utterly impossible.

But will quantum theory provide the answer to perhaps the biggest question of all: how did the universe begin? Paul Davies, a British cosmologist now at the University of Adelaide, expresses the current state of play thus: 'A few years ago I had no real answer. Today I believe we know what caused the Big Bang.'

The outsider who triggered a revolution

His optimism is based on the work, published in 1981, of a young American theorist named Alan Guth. Curiously for one of today's leading lights in cosmology, Guth began his professional life in the mid-1970s as a particle physicist at Stanford University in California. His research interests were in the implications of grand unified theories (GUTs) that bring together three of the four fundamental forces: the electromagnetic, weak and strong interactions.

One fateful day in 1978, he decided to drop in on a public lecture being delivered on cosmology by Robert Dicke, the Princeton physicist who had helped uncover the cosmic significance of the microwave noise found at Bell Laboratories in the 1960s. Guth learned of the successes of the Big Bang model and, more importantly, of its outstanding mysteries. In particular, Guth heard for the first time of the so-called 'flatness problem'.

Einstein's GR describes the idea of the force of gravitation in terms of the curvature of space and time. Just as the sun curves the space-time of the solar system, so the spacetime of the whole universe is itself shaped by the matter – the galaxies, dust and debris – it contains. GR enables the 'shape' of space to be calculated once one knows the concentration, i.e. the *density*, of matter in the universe. Until Einstein, the view was that space was, by definition, flat. Good old-fashioned geometry still applies in such space – for instance, the angles of a triangle always add up to 180 degrees. But Einstein revealed that, somewhat mind-bogglingly, gravity can cause space itself to be 'positively' curved, like a ball, or 'negatively' curved, like a saddle. What type of universe we get depends on the density and dynamics of matter within it. In particular, GR shows that at a certain critical density, the universe is *just* flat.

Astronomers have made strenuous efforts to measure the density of the universe and have arrived at a broad range of values within which the real value is expected to lie. As one might expect, the figures are not very high in everyday terms: the universe really does consist chiefly of empty space. The estimates range from the equivalent of about one proton per cubic metre to about fourteen. By comparison, one cubic metre of air contains about 10^{27} protons.

But although low, this range of values is very significant: it includes the critical density value. This can be calculated from measurements of the current rate of expansion of the universe, i.e. from Hubble's constant. And current estimates of H suggest that the critical density is about seven protons per cubic metre of space.

Listening to Dicke in 1978, Guth then heard something very interesting and strange. Using some fairly simple mathematics, Dicke showed that unless there had been some sort of interference with the universe at a very early stage, it was very hard to explain quite *why* the current density is even remotely near to the critical density. Working his calculations backwards in time, closer to the Big Bang, Dicke showed that the present small difference between the two densities now must have been the result of an *incredibly* small difference in the past. This is because the expansion of the universe vastly amplifies any differences that originally existed between the critical and actual densities.

Dicke and his longtime Princeton colleague Jim Peebles had, in fact, worked out that just one second after the Big Bang, the density of the universe must have been within one part in one thousand million million of the critical value needed to make spacetime flat – an astonishing level of fine-tuning.

Like most physicists, Guth didn't believe in someone meddling with the universe in its earliest moments and fine-tuning the densities. He believed that Nature itself must hold the answer to the flatness problem.

Returning from the lecture, Guth started to play around with the equations of cosmology. He was to discover that the answer to the enigma of the coincidence lay right in his own field of expertise: particle physics.

It takes GUTs to do cosmology

Guth was not the first to apply GUTs to the universe. Soon after the emergence of GUTs in the mid-1970s, theorists realised that these theories might have some interesting implications for the very early universe. The reason was that GUTs seek to unify three forces that are most certainly not like each other today: the electromagnetic, weak and strong forces. But according to GUTs, at very high energies – far beyond those possible in man-made particle accelerators – the weak, strong and electromagnetic forces start to 'melt' into one another and become a single force.

The energy at which the unification occurs is equivalent to a very high temperature: around 10^{27} degrees, unimaginably hotter even than the centre of the sun. Such a temperature has existed only once in the history of the universe: during its birth. Einstein's theory shows that temperatures at or above this incredible value existed during the first 10^{-34} seconds. Up until then, the three forces making up GUTs had all been one single 'GUT force'. But as the universe expanded, it cooled and as the temperature dropped below 10^{27} degrees, the GUT force broke up. First the strong force separated from the electroweak force. Then, around one hundred-millionth of a second after the Big Bang, the electroweak force split into its constituent electromagnetic and weak forces. All three forces changed their strengths at different rates so that now, fifteen billion years or so later, each is radically different in strength from the other.

All this gave the GUT theorists a new way of putting their ideas

– doomed always to be beyond the abilities of Earthbound accelerators – to the test: they began to consider the early universe as a
huge accelerator. They started talking to their cosmologist colleagues
down the hall.

By the late 1970s, some interesting results were beginning to
emerge from this unlikely confluence of the physics of the very large
and the very small. One of the first was the neat explanation of why
there is so little antimatter left in the universe. Another was the
calculation from basic principles of the temperature of the cosmic
background.

But by the time Alan Guth began work on the cosmological implications of GUTs, problems were also turning up. These stemmed
from the details of what happened when the GUT force split up into
its three constituents as the universe cooled. In essence, this schism
can be thought of as similar to the change water undergoes when it
is cooled below freezing: it turns into ice, a process known technically
as a 'phase transition'.

In winter, one can see the results of a phase transition on windows.
Different parts of the window are covered with ice crystals pointing
in different directions. The sudden changes in direction of the ice
crystals, seen where two regions meet, are known as 'defects'. In just
the same way, different regions of the universe could have cooled
through a phase transition to leave very different conditions lying
next to one another. In particular, the way in which the GUT forces
split off from one another could vary from region to region, producing various types of defects. For example, if a number of GUT regions
met at a single point, they would produce a monopole – a kind of
tiny yet extremely heavy one-poled magnet we met in the previous
chapter. If the regions met in a line, they would create a so-called
cosmic string.

This was all very exciting. But when theorists tried to estimate the
number of such defects in the universe, they made an unpleasant
discovery. The universe should be swimming in these defects – and
it isn't.

The cause of the trouble was the fact that nothing goes faster than
light. Picture two regions in the early universe, both expanding and
cooling, and thus going through their GUT phase transition. Unless
each region can 'compare notes' on how it ended up after the transition, there are bound to be some differences. These differences
result in defects at the boundary of the two regions. Now, the fastest

two regions can compare notes is at the speed of light. Yet the standard theory of the early universe predicted that many regions would have been so far apart that not even light could have travelled fast enough to minimise the creation of vast numbers of defects. Many regions would have been separated forever by the rapid expansion of the universe, without the chance to compare notes and thus avoid defects.

The end result, therefore, would be a huge number of dissimilar regions and huge numbers of defects such as magnetic monopoles. So much energy would have been locked up in these defects that they could have halted the expansion and caused a recollapse of the entire universe long before anyone got around to study it.

Before he heard Dicke's lecture, Alan Guth had, in fact, been studying the development of precisely these defects in GUTs with Henry Tye at Cornell University. After hearing Dicke, Guth hatched a bold plan. He decided he was going to use his knowledge of GUTs to see if they could solve the defects problem and the flatness problem in one go. He succeeded – and in the process produced the best theory we have of events in the very early universe.

Cosmic inflation

Guth's breakthrough was to realise that the GUT phase transition 10^{-34} seconds after the Big Bang could be of importance not only for the fundamental forces but for the whole universe as well. He discovered that it was possible that the GUT phase transition took place a little sluggishly. As a result, the splitting up of the GUT force could have been momentarily left behind by other events, leaving the universe teetering in an unstable state.

This might seem a little contrived but it is, in fact, a phenomenon that has been extensively studied in rather less esoteric circumstances. Scientists have known for years that if pure water is suddenly chilled ('supercooled') through the liquid-to-ice phase transition, the water can be cooled to 20°C below its normal freezing point before it finally undergoes the transition to ice.

Anything from a cup of water to the entire universe always tries to end up in a state of lowest possible energy. Were it not for the GUT phase transition, the expanding, cooling universe would settle down into its own lowest possible state – the so-called 'true vacuum'.

But instead, said Guth, the universe found itself stuck momentarily in an unstable 'false vacuum', at an energy far above the true vacuum.

It is this false vacuum that holds the key to Guth's new vision of the early universe. Theory reveals that its momentary presence has a dramatic effect on the dynamics of the universe. And – irony of ironies – its effects are exactly those of the lambda term, measured by the cosmological constant, which Einstein so bitterly regretted using in his first attempt to explain the dynamics of the universe.

Einstein used the lambda term to give a static universe. This required it to have an antigravitational effect to cancel out the effect of gravity. The false vacuum also produces an antigravitational effect. However, in the early universe it is so huge that it does more than counteract gravity: it vastly accelerates the expansion of the universe. Thus the false vacuum forces the universe to undergo enormous growth from an initially tiny size – according to theory about one thousand-millionth the radius of a proton – doubling its size every 10^{-34} seconds or so.

But after about 10^{-32} seconds, the unstable false vacuum collapses. The phase transition can then be completed, with the strong and electroweak forces parting company. Yet during the incredibly brief life span of the false vacuum, the universe has had time to undergo about $10^{-32}/10^{-34} = 100$ doublings in size: a factor of about 10^{30}. (The exact figure depends on very uncertain guesses concerning the particle physics at work during these times, and may be much larger.) The initially sub-sub-subatomic-sized universe has been 'inflated' up to a few inches across.

But the false vacuum has not just vanished without trace. The huge amount of energy it contained had to go somewhere, and this is where the Big Bang comes in. When it collapsed down to the true vacuum, the false vacuum poured its energy into the universe. The result was a sudden, incredibly intense burst of heat, light and radiation that filled the entire universe. Quantum effects triggered by the heat 'boiled' countless trillions of subatomic particles out of the vacuum. It was these that later went on to form atoms, molecules and – much later – the human race and everything we see.

Thus there were, in fact, two incredibly hot stages in the universe: at its creation and at the collapse of the false vacuum. Because of its direct relevance for the universe now, this second incredibly hot stage of the universe could be considered to be *the* Big Bang. The huge momentum of the inflationary phase has been enough to keep the universe expanding ever since, its effects still visible today through the red shift of distant galaxies.

The lambda term, which Einstein came to loathe so much, thus had only a brief but telling existence. Today, the universe seems to be finally in its lowest possible energy state, with zero vacuum energy. In fact, quite *why* today's vacuum is so utterly devoid of energy is a very deep mystery. Quantum theory – specifically, the uncertainty principle – dictates that even a true vacuum is seething with particles, popping in and out of existence. It is possible to calculate the energy all these particles contribute to the vacuum state today; when this is done, the answer is simply huge. Somehow, all this quantum energy is being cancelled out to give the essentially zero energy that is actually observed. Quite what is performing this trick is unknown.

Inflation solves the cosmic conundrums

Guth's cosmic breakthrough with GUT-powered inflation gives a new vision of events in the very early universe. But he actually developed his ideas in an attempt to solve two major puzzles about today's universe: what happened to all the defects formed in the early universe, and why is the universe flat?

The defects problem is easily solved. Inflation means that there was far more expansion in the early universe than the old theories predicted. Thus even parts of the universe now separated by enormous distances could once have been close together, close enough to compare notes and become like one another. So even if huge numbers of defects such as monopoles were created in the early universe, the colossal inflation diluted their numbers down to a negligible level.

The flatness problem is also quickly knocked off by Guth's ingenious idea. Think of a balloon: as air is pumped into it, it takes on a roughly spherical shape as its surface becomes curved. But as more air is pumped in, the balloon gets bigger and the surface becomes less steeply curved (see diagram on opposite page).

In just the same way, the curvature of the whole universe becomes increasingly less steep as it expands. Even though it was initially very small and very tightly curved, by the time inflation has performed its magic, the curvature is far less steep. By now, billions of years later, it is essentially flat. Guth now had an answer to Dicke's riddle about

How Guth's idea solved a cosmic mystery:
as the universe inflates, its curvature becomes progressively less pronounced

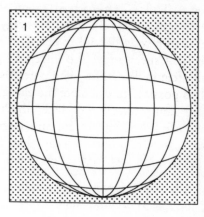

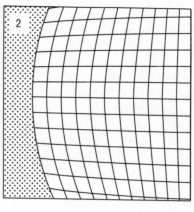

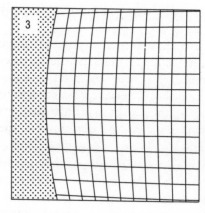

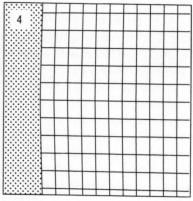

what had caused the universe to be so close to being perfectly flat today.

Yet despite all these successes, Guth's original theory suffered from a fatal flaw: the collapse of the false vacuum leads to the creation of huge 'bubbles', whose walls are packed with energy. This would lead to a far from smooth universe with a composition quite unlike the one seen today. Fortunately, by the end of 1981, a neat solution was found by another young theorist, Andrei Linde of the Soviet Union. The trick was simple: make sure that the bit of the universe we can now see was originally all part of just one of these myriad bubbles. Then the different conditions in all the others do not matter; they can never be observed.

This trick can be achieved by juggling the equations of GUTs to ensure that the collapse of the false vacuum does not occur too quickly. Then the inflation process has time to pump up one single region to form a huge volume big enough to contain the entirety of space we now see.

Two theorists at the University of Pennsylvania, Paul Steinhardt and Andreas Albrecht, came up with the same idea a few months later and the three now take the credit for inventing what is known as the 'new inflationary model'.

Since its invention, theorists have been working hard trying to get further predictions out of the theory to see if it is more than just a neat explanation of what we already know. While inflation theory's impressive ability to get around the flatness and horizon conundrums has remained essentially intact, the theory has had rather less success in correctly predicting more detailed features of the universe.

One of the first questions asked of the theory was whether it could solve the long-standing mystery of the origin of the galaxies. Recall that the key problem here is one of ensuring there is enough lumpiness in the universe to act as 'seeds' on to which gravity can pull surrounding matter to form galaxies.

Inflation theory can certainly produce lumpiness. It arises when different regions of the universe undergo transitions at slightly different times, leaving them with slightly different amounts of energy – and thus, by $E = mc^2$, mass – to act as seeds. The good news for inflation theory is that the variety of lumpiness thus generated seems to coincide nicely with that demanded by observation. The bad news is that, at least when the simplest GUT is used to explain inflation, the size of the fluctuations comes out far too large. However, as we have pointed out before, we know the simplest GUT cannot be right, so there's some hope that a more sophisticated GUT could still get inflation to crack the galaxy mystery.

One of the attractions of the earliest version of inflation was that the universe could have begun in a fairly broad range of different states and yet still have ended up like the universe we now see. Given our lack of knowledge about the state of the extremely early universe, and theorists' distaste for fine-tuning to get acceptable results, this degree of latitude is very welcome. Yet more recent versions of the theory have required such a degree of tinkering with parameters governing GUTs that one can see signs of fine-tuning creeping back in again.

In 1989, Paul Steinhardt proposed a new version of inflation that gets rid of much of the fine-tuning and, as a bonus, seems capable of giving better agreement with measurements of the density of the universe. Called extended inflation, Steinhardt's theory proposes that the rate of expansion of the universe during inflation was relatively slow. By enabling the bubbles to merge and become similar to one another, this gets over the bubble problem that killed Guth's original theory. Its advantage over Linde's original modification of Guth's idea is that it can do this without too much fine-tuning of the way the false vacuum decays.

Steinhardt has also shown that his theory makes it possible to meet the tight constraints on lumpiness, laid down by the cosmic background measurements, with a broad range of parameters, again reducing the amount of fine-tuning that has to be done. Better still, his theory does not stipulate that the universe be almost perfectly flat. As we shall see, this is a major advantage, as some astronomers think there may not be enough matter in the universe to bring about this flatness.

All the advantages of extended inflation come at a price, however: the strength of gravity must have varied during the early universe. This seems to be yet another case of theorists resorting to a *deus ex machina* to get them out of a corner. But it is not so wild an idea. Virtually all particle theories, including superstring theories, allow the possibility that some energy field be linked to the strength of gravity in the universe. In any case, gravity doesn't have to vary for long: as soon as the inflationary phase ends, the link with the energy field is severed and gravity keeps a steady value ever after.

Steinhardt's latest version of inflation seems destined to stay around for some time yet, as other theorists study its equations looking for more predictions it might make. But all theories of cosmic inflation deal only with the universe very shortly after its birth and not with the moment of birth itself. On that event itself, GUTs have nothing whatever to say. And so we come to the deepest question of all: how did the universe actually begin?

In the beginning

Guth's concept of inflation, with its GUT-fired reheating of the universe, may give a new meaning to the Big Bang but such a redefinition would strike many as a cop-out. The central question remains: what triggered the creation event itself?

The implication of Hawking and Penrose's singularity theorems seems to be that the only thing we know for sure about the instant of creation is that we can never know anything about it. At a singularity, many of the parameters we need to describe the universe, such as its density and temperature, take on ludicrous, infinite values. But a glance at the premises on which the singularity theorems are based offers the possibility of a solution for this ultimate problem. Two assumptions, in particular, attract attention. The first is that a GR-type theory of gravity is valid. The second is that matter in the universe always behaves, roughly speaking, in a straightforward fashion. For example, we cannot have any antigravitational effects.

Both these premises can be circumvented by introducing quantum effects. Einstein's theory of gravity may be the best one we have at present but it reflects Einstein's ultimate distaste for quantum ideas and contains no quantum ideas at all. Put such ideas in, and premise number one might go. We have also seen how quantum physics can produce very weird behaviour quite unlike that seen in ordinary matter: the antigravitational effect of the false vacuum state is an excellent example. Thus might premise number two fall.

Unfortunately, as pointed out earlier, we still don't have a complete theory of quantum gravity. Even so, a few things can be said about the effects quantum gravity would have on the very early universe. First, we can estimate the length of time such effects played a key role in the behaviour of the universe. General arguments show that they must have dominated the first 10^{-43} seconds after the universe burst into existence.

Professor Stephen Hawking and colleagues at Cambridge University have been concentrating their efforts on understanding what happened during these very first instants after the creation. In particular, they are looking at how the very nature of space and time itself is turned upside down by quantum effects.

Calculations by Hawking and his team suggest that during the brief reign of quantum gravity, effects resulting from the uncertainty principle blurred the distinction between space and time. It may then, they argue, be possible to think of time as just another spatial dimension. This loss of special identity for the time dimension then makes possible a rather interesting trick. Imagine an ant crawling around on the surface of a ball. Give the ant long enough and it will explore all of the ball's surface – the surface area is, after all, finite. But no matter how long the ant spends crawling around, it never finds an *end* to the surface of the ball. This simple example illustrates something of great importance in Hawking's work: it is possible to have a geometrical figure which is finite in extent and which yet has no boundary.

It turns out that quantum gravity allows exactly the same situation to occur with spacetime: it can be finite in extent and yet have no boundary. Thus the universe could have an origin (i.e. a finite extent in spacetime) and yet have no starting point in time or space! The question of what happened at (or before) the moment of creation then becomes meaningless. Hawking explains: 'Time ceases to be well defined in the very early universe, just as the direction "north" ceases to be well defined at the North Pole. Asking what happens before the Big Bang is like asking for a point one mile north of the North Pole.'

But the most controversial aspect of the no-boundary idea is that it makes the concept of a creator redundant. As Hawking somewhat bluntly puts it: 'If the universe is really self-contained, having no boundary or edge, it would have neither beginning nor end: it would simply be. What place, then, for a creator?'

Many might find this final conclusion disturbing, even offensive. It may be that the no-boundary condition does not, in fact, apply to our universe – Hawking himself stresses that it is only a proposal. However, it has to be admitted that Hawking's is not the only theory to do away with any obvious need for a creator.

Was the universe born from nothing?

The no-boundary condition is undoubtedly an ingenious approach to the profound problem of the creation of the universe. It also, according to Hawking's latest research, leads to conclusions broadly in line with observations. However, it does have the air of a mathematical sleight of hand about it and a number of theorists are taking a rather different, more 'physical' approach to the great mystery of the creation.

Their starting point is, once again, quantum theory. One consequence of Heisenberg's uncertainty principle is that it is possible to create – literally from nothing – a certain amount of energy. The only demand is that the energy does not exist for too long; Heisenberg's equation specifies just how long a given amount of energy can exist before it must vanish again. According to Einstein's famous equation, an amount of energy, E, can be converted into an equivalent mass, M, via the relationship $E = mc^2$. Thus, to say that energy fluctuations can suddenly emerge from nowhere is equivalent to saying that matter can suddenly emerge from nothing.

Incredible as it may seem, all around us, all the time, countless trillions of particles are bursting into existence and disappearing again lest they contravene Heisenberg's equation. These particles are actually formed in pairs – particles and their antiparticles – and, as they burst out of seeming nothingness, they are called 'vacuum fluctuations'.

Quantum theory pulls many surprises but this one seems particularly wild. Yet the effect of these vacuum fluctuations has actually been measured in the laboratory. In 1947, Willis Lamb and Robert Retherford in America showed that the particle-antiparticle pairs produced a small but detectable change in the spectrum of hydrogen, and that the size of this change agrees with theory (based on quantum electrodynamics, QED) to better than one part in a billion.

Thus it is now known there is no such thing as 'nothing'. Even empty space is seething with these utterly random, causeless vacuum fluctuations. A small coterie of theorists in America, Europe and India has been working for some years on the possibility that, like these particle pairs, the universe itself just flipped into existence as a vacuum fluctuation. The smaller the universe, the larger these fluctuations would be – and the wilder its original behaviour. Cosmic inflation demands that the universe was originally extremely hot and expanding, and it seems likely that quantum fluctuations could provide both the heat and the antigravitational force needed to create such conditions.

If these ideas are even vaguely along the right lines, the universe would never need 'outside help' to come into being. It simply burst into existence, quite spontaneously, out of nothing. One American theorist, Edward Tryon, takes a particularly strong line on this notion. He believes that it may be that the universe itself – despite appearances – actually contains no energy at all. Heisenberg's uncertainty principle demands only that finite amounts of energy cannot last forever; Tryon's view would mean that the no-energy universe *could* last forever.

Tryon's argument exploits the fact that gravity always gets stronger the closer two objects are to one another. Technically, this means that gravity has *negative* energy. Tryon realised that it may be possible that all the negative gravitational energy of the universe is exactly balanced by all the positive energy in the form of particles. It turns out that in our universe these two forms of matter are very nearly – perhaps exactly – equal. Thus the *total* energy in the universe may be precisely zero – in which case we are all now simply sitting in a huge, causeless, expanding vacuum fluctuation.

There are some technical problems with Tryon's theory but growing numbers of theorists are taking his ideas seriously. If correct, the theory would lead to a rather prosaic view of the creation, an event which has lain at the centre of countless philosophical and religious debates for centuries. It would imply that, as Alan Guth, father of inflation, puts it: 'The universe may be the ultimate free lunch.'

How will the universe end?

The question of how the universe began is, as we have seen, more or less entirely in the hands of the theorists. The huge amount of inflation and heat generated by the primordial phase transition wipes out everything that went before, making the events right at the very beginning forever beyond our reach.

The question of how the universe will end is, however, still very much in the court of the traditional astronomer. But, as we shall learn, particle theorists are increasingly getting in on the act here as well. The answer to this second cosmic conundrum may one day be provided by a particle accelerator rather than a telescope.

How the universe will end depends on whether or not the gravitational forces within it are strong enough to halt the present expansion. A simple analogy explains the situation. Think of a ball being thrown into the air. The harder you throw it, the higher it goes before gravity stops its upward flight and brings it back again. Throw the ball hard enough, though, and the ball will never return. It will be moving so fast that gravity cannot stop it.

Thus what happens to the ball depends on the strength of gravity. And so it is with the entire universe. Calculations show that if the density of matter in the universe exceeds a critical level of an average of about seven protons per cubic metre of space, the current expansion will come to a grinding halt and then reverse. Gravity will suck the universe inexorably back into a cataclysmic Big Crunch some tens of billions of years hence.

In view of the awesome events that would take place if we *were* in a universe at the critical density, it is probably not surprising that many astronomers are deeply interested in the question of whether or not the amount of matter in the universe exceeds this critical density value. The ratio of the actual density of the universe to the critical value has been dubbed Omega by theorists – an appropriately eschatological name.

Thus if Omega is less than one, the universe will happily expand forever. But should Omega be greater than one, our extremely distant descendants are in for a tough time.

On the face of it, finding out the density of matter in the universe doesn't seem that difficult, just rather tedious. The universe is full of galaxies separated by huge empty space. We can estimate the number of galaxies in the visible universe simply by counting those in a suitably huge block of space and assuming that the resulting number of galaxies per million cubic light years is typical of all space.

The next step would be to work out how many of these blocks there are in the visible universe; this gives the total number of galaxies within it. Finally, we would need to work out the mass of each of these galaxies. To do this, we need to know the number of stars each contains. This again can be estimated by taking suitably big samples. Assume for simplicity that the average star in a galaxy is like our sun and has a similar mass.

We now have all we need to estimate the mass of the universe.

Multiply the number of galaxies in the visible universe by the number of stars in a typical galaxy – giving the number of stars in the visible universe – and finally multiply this by the mass of the sun.

This can, in fact all be done. The figure that emerges for the mass of the visible universe is suitably huge: about a million billion billion billion billion billion (10^{51}) kilograms. To get the all-important density, however, we must divide this figure by the volume of the universe. This can be calculated from a knowledge of Hubble's constant, H, which determines how fast the universe is expanding.

When we do this, we finally get a figure for the density of the universe of about 0.1 protons per cubic metre. To gauge the significance of this figure, we divide it by the critical density. The value of this, it should be emphasised, depends quite sensitively on the value of Hubble's constant, H. As the precise value of H is still a matter of some contention among astronomers, there is some doubt over the critical density figure. However, the current consensus is that the critical density is equivalent to around seven protons per cubic metre.

Dividing this into the figure of 0.1 protons per cubic metre for the *actual* density of matter in the universe thus gives a value for Omega far below the dreaded value of 1.

This little calculation suggests we have nothing to fear from the future. There just isn't enough matter in the universe to halt the expansion. Unfortunately, some disturbing research carried out in the 1930s by an astronomer at the California Institute of Technology shows that we may still be in trouble.

The discovery of dark matter

Fritz Zwicky, born in Bulgaria of Swiss parents in 1898, came to Caltech in 1927 determined to find out more about supernovae, the incredibly violent death throes of stars. But by the 1930s, the rarity of these events (only two have been seen in our own galaxy in the last 400 years) led him to turn to more immediately satisfying work on galaxies.

Here, his research quickly bore fruit. He discovered that galaxies tend to form in clusters, each one typically containing 100 or more galaxies. To find out what was going on within these clusters, Zwicky turned to a spectroscope, which reveals the movements of galaxies by the apparent shift in the spectral lines of their light.

This revealed that the galaxies within one cluster in the constellation Coma Berenices were moving rather fast relative to one another. But when Zwicky estimated the amount of matter in the cluster, he found something odd: there didn't seem to be enough matter to prevent these fast-moving galaxies from racing out of the cluster.

Zwicky realised the answer: there must be more matter in the cluster than can be picked up by conventional optical telescopes. These can see only matter giving out (or reflecting) light. Clearly, there must be dark gas, dust, dead stars and who knows what else lurking in the cluster.

Zwicky had thus discovered the presence of the now infamous 'dark matter' which holds the key to the fate of the universe. More than half a century after Zwicky warned of its existence, finding out what it comprises and how much of it there is in the universe remains one of the prime research targets of astronomers.

Careful measurements of the movements of star clusters about our own galaxy have now revealed the presence of huge amounts of dark matter. It is thought that there is more than ten times more dark matter in the galaxy than there is of the visible matter we see through telescopes. It appears to stretch far beyond the apparent edge of our galaxy, forming a huge spherical halo around it.

Studies of the behaviour of other galaxies nearby all reach more or less the same conclusion: for every kilogram of luminous visible matter in the universe, there are about ten kilograms of dust, gas and other dark matter.

A completely independent way of checking this result comes from the way in which some chemical elements were made in the furnace of the Big Bang. As we learned earlier, only relatively light elements can be formed in the Big Bang. The link with dark matter comes from the fact that the abundance of these elements depends on the density of protons and neutrons – that is, baryons – that were available at the time. In particular, the amounts of helium-3, an isotope of helium, and deuterium, an isotope of hydrogen, that exist today give a way of estimating the density of 'ordinary', i.e. baryonic, matter in the universe. Measurements of the abundance of these two isotopes in the cosmos give a figure for the overall density in line with that derived from the observations of galaxies.

All these observations lead to a new estimate of the true density of the universe. When baryonic dark matter is taken into account, the figure goes up from about 0.1 to 1 proton per cubic metre. Fortunately, however, the result still gives an Omega value well short of 1.

So are we finally safe from having to face a Big Crunch? Not yet. As particle physicists know only too well, baryons are not the only type of particle in existence. There could be non-baryonic material quite invisible to telescopes out there in space, bringing the universal expansion to a grinding halt.

The search for the particles of doom

The first species of particle to be considered as the particles of doom was a rather unlikely candidate: the neutrino. These non-baryonic particles were dreamt up to explain some curious features of radioactive decay and were long thought to have no mass at all. They don't interact very strongly with ordinary matter – they could pass through a million billion kilometres of rock without a hitch. All in all, neutrinos are not what one would think of as very promising candidates for halting the expansion of the universe.

Yet theorists discovered that if neutrinos were fast-moving (i.e. hot) and had even a very small amount of mass – as little as 0.0005 of that of the electron – their huge numbers throughout the cosmos could result in a high enough density to produce an Omega value of 1.

So when, in 1980, a team of Soviet experimentalists based in Moscow announced that they had found evidence that the neutrino did have rather a high mass after all, the effect on astronomers and particle physicists alike was electric.

The experiment needed to find the mass of the neutrino is, however, a very delicate one. When other groups tried to repeat the Soviet findings, they had no success. They did, however, put an upper limit on how heavy the neutrino could be. This was about half the mass found by the Soviet experiment, although a value of just zero could not be ruled out. Ten years later, a joint Soviet and American research team reported results of measurements of the flow of neutrinos out of our own sun that implied that the neutrino does, in fact, have a mass. But this time the implied value is so low (about two billionths the mass of the electron) that it has no cosmological importance.

But by then, the idea that *any* form of hot, dark matter, such as fast-moving, massive neutrinos, could halt the expansion had run into an even more basic problem: a universe dominated by such dark matter would not look like the one we live in.

This discovery was made by astrophysicists using not telescopes but supercomputers. They programmed the computers to simulate the effect of neutrinos and ordinary matter in model universes, set them running and watched what happened. It soon emerged that the model universes dominated by neutrinos were too clumpy. Ludicrously huge islands of matter appeared, surrounded by similarly vast areas of empty space. Astronomers know, however, that in the real universe matter is much more smoothly distributed.

So, for the present, invisible, fast-moving particles have fallen by the wayside as candidates for the particles of doom. Needless to say, theorists have managed to put up new candidate particles in their place. One lesson has been learnt from the defunct idea: to stand a chance of being the real particles of doom, any candidates must not whizz around the universe so quickly that they obliterate clumps of matter trying to form into galaxies.

The generic name for slow-moving particles with a chance of meeting this test is 'cold dark matter', or CDM. A number of plausible candidates for CDM have been put forward in recent years, and they all have one thing in common: no one knows if they exist. We have, in fact, already briefly met two CDM candidates: magnetic monopoles and cosmic strings. Both are expected to be born during the GUT phase transition that propelled the cosmic inflation. In particular, each would form at the boundaries between regions in which that phase transition turned out rather differently. If the regions meet at a point, the result would be a pointlike monopole. If, however, the regions meet at a line, the result would be a cosmic string.

There have been occasional reports of both these weird types of objects, but thus far none has been widely accepted by astrophysicists.

Particle physicists have not missed out on their chance to put forward candidates for CDM. Arguably their best candidate so far is the rather inappropriately named WIMP – the Weakly Interacting Massive Particle. It seems tailor-made: it is massive, so it can do the job, but it also interacts only weakly with ordinary matter, so we won't have seen it yet.

But WIMPs aren't pure fancy. They emerge from supersymmetry, one of the most important ideas of particle physics. This powerful principle implies that the particles of matter have supersymmetric partners with weird names like squarks (for the partners of quark), and selectrons. It also leads to a prediction for the mass of the lightest of these supersymmetric partners (LSPs) for ordinary matter. Intriguingly, this lowest mass turns out still to be high enough to halt the expansion of the universe. Thus, being both weakly interacting and quite massive, LSPs are prime candidates for WIMPs, and WIMPs would make great CDMs!

The possibility that WIMPs are still out there, floating through the cosmos, has led to the setting-up of a number of ingenious experiments to detect them.

Scientists at a number of British universities are attempting to detect the presence of WIMPs by exploiting the phenomenon of superconductivity.

When some metals are cooled to close to absolute zero, $-273°C$, electrons within them team up into pairs (called Cooper pairs) which suffer fewer interruptions to their passage through the metal. In fact, the Cooper pairs flow more or less unimpeded through the metal, which thus loses all its electrical resistance and becomes a 'superconductor'. The electrons forming a Cooper pair are, however, bound only weakly to one another and can be relatively easily split up. Scientists at Oxford University and elsewhere are hoping to exploit this fact and to build a device which detects the breaking apart of Cooper pairs caused by the impact of a WIMP.

Huge particle accelerators, usually thought of as probes of the very smallest regions of space, are also joining the hunt for CDM. So far, the results have been somewhat negative: data from the Stanford Linear Accelerator Center in California and LEP at CERN in Geneva have recently been able to rule out some ideas for WIMPs. They have also restricted the amount of leeway left to theorists wanting to make supersymmetric partners of ordinary matter the particles of doom.

This weeding-out of the no-hopers for CDM is making life tough for the theorists. As John Ellis, leader of the theory division at CERN, put it recently: 'LEP has made it more difficult to find out what dark matter is.'

But perhaps the struggle the theorists are having has a very simple explanation: perhaps Omega really *is* less than 1, and there really isn't enough matter in the universe to cause a Big Crunch. There are no laws *demanding* a recollapse – although not often mentioned even by experts, even inflation can accommodate a value of Omega other than 1. So what would happen if the universe simply kept on expanding forever?

The long, slow death

Scientists were speculating on the ultimate fate of the universe long before anyone knew of the cosmic expansion. In 1854, the German physicist Hermann von Helmholtz used the laws of thermodynamics to suggest that the universe was heading inexorably toward so-called 'heat death'. All the energy that drives change in the universe is being used up, he argued, and so eventually all matter will gather into a ball of uniform temperature. Popularised by some of Britain's leading astronomers in the 1930s, the vision of heat death prompted some leading figures of the day to ponder glumly about this dismal end to all things.

Technical arguments based on Einstein's theory of gravity rule out a Helmholtz-style heat death from occurring in an expanding universe. Even so, a fate just about as dismal could still be on the cosmic agenda.

As far as we on Earth are concerned, the end will come with the death of the sun – unless, of course, we have found a way of leaving the solar system and thus buying time for ourselves. Our sun is expected to start running out of nuclear fuel in about 5,000 million years. Being a relatively small star, its end will not be a spectacular supernova explosion. As its fuel runs out, the sun is expected to balloon outward to form a so-called red giant star such as the well-known star Betelgeuse in the constellation Orion. The sun will end up something like 800 times larger than it is now – big enough to engulf the orbits of the first four planets, including the Earth. It will spend its final millennia gently blowing off its outer layers in vast puffs of particles, eventually leaving behind just a tiny white dwarf star no bigger than the Earth itself.

If we have succeeded in travelling to the stars by the time these events come to pass, we shall have bought a very substantial amount of extra time. The rate at which stars are being born is declining, simply because the matter born in the Big Bang is used up. Even so, there is enough raw material left to support the birth of stars for another thousand *billion* (10^{12}) years or so. During this time, we will be free to hop from one to another.

But when this time is up, there will be nowhere left to go. As their component stars are snuffed out and gravitation energy leeches away, galaxies will start to fall apart in about 10^{18} years. The universe will then be filled with the remnants of dead stars. Most of these will be black dwarfs – the husks of stars like our sun – floating out of the dismembered galaxies.

But some other stars will have turned into altogether more bizarre objects: black holes. When a star about ten times the mass of the sun runs out of fuel, there is no known force capable of stopping gravity from causing the complete collapse of the dying star. As the collapse continues, the density of the object and the strength of its gravity field soars. Eventually its gravity becomes so intense that not even light can escape its clutches. A black hole has been formed.

Anything venturing into a black hole would suffer an appalling fate. So intense is the gravity field that an astronaut would be simultaneously crushed and torn apart by the difference in the level of gravity acting on his feet compared to his head.

So, as the galaxies disintegrate, huge black holes would clump together in space, sucking in whatever lay within their grasp. After countless aeons, however, even these black holes would disintegrate, falling prey to a phenomenon discovered by Stephen Hawking in 1974. By applying quantum theory to the physics of the black hole, he discovered that they could give the impression of emitting particles: as Hawking puts it, 'Black holes ain't so black.' As the emission continues, the black hole gets hotter and hotter until eventually it evaporates in a burst of intense radiation. Black holes of the size of the sun and larger would emit particles extremely slowly. Even so, the Hawking process still applies. After the unimaginably long time of 10^{122} years (that's 1 with 122 noughts after it), even the largest black holes in the universe will have evaporated.

The *absolute* end of everything will come after an amount of time that makes even these previous timescales seem like the blinking of an eye. If current grand unified theories are correct, the protons that make up all atoms will have long since vanished. But even if GUTs are wrong, all matter will eventually just vanish in a puff of radiation. About 10 to the power 10^{26} years from now, iron – the most stable of all atoms – will collapse into tiny black holes as the result of so-called quantum tunnelling. These black holes will then briefly flare up in one last tribute to the work of a long-dead Cambridge scientist.

Nothing at all will be left in the universe. It will be – The End.

Further Reading

This is not a scholarly book, and what follows is not intended to be a comprehensive list of references for every fact I cite. What the list below is intended to provide is an entrée into the more detailed literature that exists in the various fields covered by the book, so that the interested reader can pursue his or her fancy in greater depth.

General

The latest progress on the Outstanding Mysteries investigated here is usually announced in academic journals such as *Nature* and *Science*. These are, however, often daunting reading. The best way of finding out about the latest advances is to read the science pages of broadsheet newspapers (mostly written by scientifically qualified staff) and magazines such as *New Scientist*, *Scientific American* and *Discover*.

Chapter 2

- Francis Crick, *What Mad Pursuit* (Weidenfeld & Nicholson, 1988). Crick has not rested on his laurels after his work on DNA. This book provides an intriguing insight into the thoughts and personality of one of the world's most creative scientists.
- Richard Dawkins, *The Blind Watchmaker* (Longman, 1986). Addresses in a vivid way the apparently simple but very deep questions surrounding the genetic basis of evolution.
- Renato Dulbecco, *The Design of Life* (Yale University Press, 1987). A comprehensive yet accessible overview of the science of life, covering everything from growth to designer drugs. Written by a

Nobel prizewinner, it can be tough in parts but not half as hard as the subject itself.

- John Gribbin, *In Search of the Double Helix* (Corgi, 1988). A clear account of the long trail leading from Darwin's theory of evolution to the discovery of the role of DNA in genetics and beyond. Gives much more detail on the chemistry underlying molecular biology, and on the complexities of the life process.
- Mark Ridley, *The Problems of Evolution* (Oxford University Press, 1985). Gives answers (or as much of an answer as is currently possible) to some of the most frequently asked questions about evolution theory. Very clearly written.
- Robert Shapiro, *Origins* (Pelican, 1986). A reasonably up-to-date account of research into the ultimate origins of life, written in an entertaining, witty and acerbic style by a leading chemist who takes a lot of accepted wisdom with a pinch of salt.
- James Watson, *The Double Helix* (Penguin, 1974). Arguably the most famous account of science in action ever written. The story of the discovery of the secret of DNA is recounted by one of the Nobel prizewinners involved in enough personal detail to have caused a scandal when first published in 1968.

Chapter 3

- Preston Cloud, *Oasis in Space* (Norton, 1988). Detailed and up-to-date account of Earth science, written by one of its leading lights.
- Victor Clube and William Napier, *The Cosmic Winter* (Blackwells, 1990). Written by two British astrophysicists, this disturbing book should shake anyone out of complacency about how safe our planet is from cosmic catastrophe.
- James Gleick, *Chaos: Making a New Science* (Viking, 1987). Lively introduction, shorn of mathematics, to the ideas and personalities behind chaos theory.
- John Houghton, Geoff Jenkins and Jim Ephraums, *Climate Change: The IPCC Scientific Assessment* (Cambridge University Press, 1990). A detailed, technical and independent review of the state of knowledge on the condition and likely future of our climate, prepared for the United Nations Intergovernmental Panel on Climate Change. Useful for deflating optimists.

- Richard Muller, *Nemesis: the Death Star* (Heinemann, 1988). Written by a distinguished astrophysicist who worked with the Alvarezes on uncovering the K/T extinction event, this shows clearly how science really works – with back-of-the-envelope estimates, races to get into print before rivals and all the other aspects of scientific life scientists don't like to talk about.
- Fred Pearce, *Climate and Man* (Vision Books, 1989). Non-technical guide through some very technical ideas on the impact of climatic change on life, including the social and economic issues involved.
- Ian Stewart, *Does God Play Dice?* (Basil Blackwell, 1989). Better than Gleick for getting a genuine feel for chaos and its applications, but includes some (basic) mathematics.

Chapter 4

- David Bohm, *Wholeness and the Implicate Order* (Ark, 1983). A difficult book, especially for the non-mathematical, about a difficult subject: is the current view of quantum theory the last word? Bohm is one of the few dissenters.
- Paul Davies and Julian Brown (Eds.), *The Ghost in the Atom* (Cambridge University Press, 1988). A series of interviews with many of the authorities on what quantum theory actually means. They make clear that little is clear. Worth getting for the introduction alone.
- Robert Eisberg, *Fundamentals of Modern Physics* (Wiley, 1961). If you can take mathematics at A-level, you should find this illuminating on many issues of basic quantum theory.
- Paul Matthews, *Introduction to Quantum Mechanics* (McGraw-Hill, 1974). Remarkably clear and comprehensive guide to basic quantum theory, considering it demands only A-level mathematics.
- Abraham Pais, *Subtle Is the Lord . . .* (Oxford University Press, 1982). Definitive one-volume biography of Einstein by a theoretical physicist who knew the man. Skippable mathematics, enjoyable insights and anecdotes.
- Jonathan Powers, *Philosophy and the New Physics* (Methuen, 1982). Deals with the conceptual and philosophical questions raised by quantum theory (and other areas of modern physics) in a straightforward and clear way.

Chapter 5

- John Barrow, *Theories of Everything* (Oxford University Press, 1991). A well-argued rejoinder to claims by some physicists that the end (i.e. a Theory of Everything) is not only possible, but quite possibly already here (in the form of superstring theory).
- Peter Coveney and Roger Highfield, *The Arrow of Time* (W. H. Allen, 1990). An immensely readable attempt by the authors to solve the mysteries of time by investigating the physics of time — from Newtonian physics to relativity and quantum physics — and its wider manifestations. Shows how something 'obvious' can have many hidden depths. Highly recommended for the general reader.
- Gale Christianson, *In the Presence of the Creator* (Free Press, 1984). Engaging account of the life and work of Sir Isaac Newton which provides ample evidence of Bronowski's contention that the ascent of man has not been made by lovable people.
- Paul Davies, *Superforce* (Unwin, 1985). Now a little dated, this is nonetheless a very clear account of quantum field theory for a lay audience. Includes much on the cosmological aspects, including inflation.
- Paul Davies (Ed.), *The New Physics* (Cambridge University Press, 1989). Outstanding collection of essays on the whole of modern physics written by world authorities. Some chapters less accessible than others, but mainly *Scientific American* level.
- Paul Davies and Julian Brown (Eds.), *Superstrings: a Theory of Everything?* (Cambridge University Press, 1988). A collection of interviews with the founders and critics of superstring theory. Davies asks them the questions we all want to put.
- Freeman Dyson, *Disturbing the Universe* (Pan, 1981). Fascinating collection of essays by a scientist who has unimpeachable credentials, yet is not afraid to consider quirky ideas.
- Richard Feynman, *QED: The Strange Theory of Light and Matter* (Princeton University Press, 1988). A characteristically idiosyncratic account by one of its originators of the basis of the most successful theory ever devised.
- Sheldon Glashow, *The Charm of Physics* (Simon & Schuster, 1991). A collection of essays on a wide range of subjects, from science education to superstring theory, by a Nobel laureate in physics.
- Richard Rhodes, *The Making of the Atomic Bomb* (Penguin,

1988). Pulitzer prizewinning account of how the founders of particle physics used their knowledge and expertise to devastating effect.

Chapter 6

- John Barrow and Frank Tipler, *The Anthropic Cosmological Principle* (Oxford University Press, 1986). A comprehensive survey of the evidence for a connection between life on Earth and the nature of the universe, by two leading cosmologists.
- Michael Berry, *Principles of Cosmology and Gravitation* (Cambridge University Press, 1976). An introduction to the mathematics of Einstein's theory of gravity and its applications to the universe, requiring only A-level mathematics.
- Martin Harwit, *Cosmic Discovery* (Harvester, 1981). Intriguing account of how we know what we do about the universe.
- Stephen Hawking, *A Brief History of Time* (Bantam, 1988). The most famous popularisation of science of recent times but decidedly hard going in places, even if you know what Hawking is talking about. However, a valuable guide to the current interests of one of our leading theoretical physicists.
- Lawrence Krauss, *The Fifth Essence* (Radius, 1990). A detailed account of the mystery of the dark matter in the universe by an expert in the field. Despite being non-mathematical, some parts are rather technical for a lay reader.

Index

Index

Index

Index